AF517837

ENCYCLOPAEDIA OF ENGINEERING

ENCYCLOPAEDIA OF ENGINEERING

VOL. 1

CIVIL ENGINEERING A-D

By

A.M. Kharmawphlang

A. Khongphai

AKANSHA PUBLISHING HOUSE

NEW DELHI

AKANSHA PUBLISHING HOUSE
4649-B/21, Ansari Road
Darya Ganj, New Delhi-110 002
Email: ektabooks@yahoo.com
Ph.: 23263193/9811582579
Fax: 011-23263193

Encyclopaedia of Engineering

First Published 2006

ISBN 81-8370-048-9 (Set)

PRINTED IN INDIA

Published by M.P. Misra for Akansha Publishing House, New Delhi and Printed at Aryan Enterprise, Delhi.

PREFACE

Engineering is the application of science to the needs of humanity. This is accomplished through the application of knowledge, mathematics, and practical experience to the design of useful objects or processes.

Engineering involves the knowledge of mathematical and natural sciences (biological and physical) gained by study, experience, and practice, applied with judgement and creativity to develop ways to utilize the materials and forces of nature for the benefit of mankind.

Encyclopaedia of engineering contains engineering terms with clear and concise definitions. The encyclopaedia is highly illustrated and written in plain, non-technical language and extensive cross-referencing makes concepts more easily understood. Timings and Twigg focus on a wide range of engineering disciplines — engineering science, electrical and electronic engineering, workshop practices and mechanical engineering — and provide definitions that are relevant to all.

This vital reference offers a wealth of essential information in a convenient, quick-find format. Whether you're a professional, a student, a writer, or a general reader with an interest in science, there is no better or more authoritative way to stay up-to-speed with the current language of the various branches of engineering or gain an understanding of their key ideas and concepts.

Encyclopaedia of Engineering is written specifically for further education students by experienced authors who have taught and written extensively at this level.

A

A Jointing. See Joining.

AAC. Autoclaved Aerated Concrete.

Abacus. The flat slab on top of a column capital, supporting the architrave.

Abandonment. The failure of both parties to a contract to abide by its terms.

Abattoir. A slaughterhouse; a place where animals are butchered for food.

Aberration. A distortion of an image produced by a faulty lens or mirror.

Above Ground Tank. A large above ground vessel used for the storage of liquids.

Abrade. Scrape or wear away by friction.

Abrams' Law. A rule stating that with given concrete materials and conditions of test, the ratio of the amount of water to the amount of the cement in the mixture determines the strength of the concrete, provided the mixture is of a workable consistency.

Abrasion Resistance. Ability of a surface to resist being worn away by rubbing and friction.

Abrasion. Wearing away by friction.

Abrasive. A substance used for wearing, grinding, cleaning, or polishing by rubbing or grinding.

Abrasive Coatings. In closed coating of paper no adhesive is exposed, as surface of paper is completely covered with abrasive; in open coating, surface of backing paper is covered with regulated amount of abrasive, exposing the adhesive; space between the abrasive grains reduces loading and filling when sanding gummy or soft materials.

Abrasive Paper. Paper with an abrasive surface; sandpaper, emery paper; garnet paper.

Abrasive Surface Tile. Floor tile that has been roughened to be slip-resistant.

Abrasive Surface. A surface that has been roughened for safety or for warning.

ABS. Acrylonitrile Butadiene Styrene; a plastic used for piping; has high resistance to impact, heat, and chemicals.

ABS Pipe. A plastic pipe sold in 10 and 20 foot lengths in various diameters for plumbing stacks and drains; used primarily for drain lines.

Absolute Humidity. The density of water vapor per unit volume of air.

Absolute Pressure. The pressure measured by a gauge plus a correction for the effect of air pressure on the gauge (14.7 psi at sea level).

Absolute Temperature. Temperature measured from absolute zero.

Absolute Viscosity. A method of measuring viscosity using the poise as the basic measurement unit; this method utilizes a partial vacuum to induce flow in the viscometer.

Absolute Volume. The volume of an ingredient in its solid state, without voids between individual pieces or particles; in the case of fluids, the cubic content occupied; in concrete, it is the actual volume occupied by the different ingredients determined by dividing the weight of each ingredient in pounds by its specific gravity times the weight of one cubic foot of water in pounds; example, the absolute volume of one sack of cement equals 94 divided by 3.15 times 62.4 equals 0.478 cubic feet.

Absolute Zero. A theoretical lowest possible temperature, at which all molecular motion ceases, calculated to be exactly minus 273.15° Centigrade or minus 459.67° Fahrenheit.

Absorb. To swallow up or suck in, like wood absorbing a finishing material.

Absorbed Moisture. Moisture that has entered a solid material by absorption and has physical properties not substantially different from ordinary water at the same temperature and pressure; in aggregates, that water which is not available to become part of the mixing water.

Absorbent. Having the ability to suck up liquid, gas, or heat.

Absorber. The blackened surface in a solar collector that absorbs the solar radiation and converts it to heat energy.

Absorptance. The ratio of light absorbed by a material to incident light falling on it.

Absorption. The relationship of the weight of the water absorbed by a material specimen subjected to prescribed immersion procedure, to the weight of the dry specimen, expressed in percent.

Absorption Chiller. A system similar to a vapor compression chiller with the exception that it does not use a compressor, but uses thermal energy (low pressure steam, hot water, or other hot liquids) to produce the cooling effect.

Absorption Coefficient. The absorption coefficient of a material or sound-absorbing device is the ratio of the sound absorbed to the sound incident on the material or device; the sound absorbed by a material or device is usually taken as the sound energy incident on the surface minus the sound energy reflected.

Absorption Rate. (1) The speed at which the real estate market can absorb new offerings of land or buildings during a specified period of time. (2) The amount of water absorbed when a brick is partially immersed for one minute; usually expressed in either grams or ounces per minute per 30 sq. in; also called suction or initial rate of absorption.

Absorption Refrigerator. Refrigerator which creates low temperatures by using the cooling effect formed when a refrigerant is absorbed by chemical substance.

Absorption, Total. The amount of water a masonry unit will absorb when immersed in water.

Abstract of Title. A written summary of all transactions that could affect the ownership of a piece of real property, including deeds, leases, liens, and wills.

Abutment. The lateral supporting structure of an arch, bridge or similar pressure; that part of a pier or wall from which an arch springs, specifically the support at either end of an arch, beam or bridge; that part of a structure which takes the thrust of a beam, arch, vault, truss or girder; the part of a bridge that supports the end of the span and prevents the bank from sliding under it; a foundation that carries gravity and also thrust loads.

Abutment Piece or Member. The bearing plate or piece of a wall system to which the loads are transferred.

AC. (1) Air Conditioning. (2) Asphaltic Concrete. (3) Alternating Current.

AC Relay. An electromagnetic or electromechanical valve or switch using small currents and voltages to control the making or breaking of electrical contacts on a circuit.

Accelerate. To quicken or hasten the natural progress of certain actions or events.

Accelerated Depreciation. The declining balance and sum of the year's digits method which give greater depreciation to the early years of the life of the assets.

Accelerating Admixture. Substance that increases the rate of hydration, shortens setting time, or increases strength development in concrete, mortar, grout, or plaster.

Acceleration. 1. Requiring change order work to be done without extension of the contract time. 2. An increase in the rate of speed.

Acceleration Clause. A clause in a mortgage or trust deed that would allow the lender to call the whole loan due at any time that certain specified events occur, such as a default in payments or sale of the property.

Accelerator. A substance which, when added to concrete, mortar, grout, or plaster, increases the rate of hydration of the hydraulic cement, shortens the time of setting, or increases the rate of hardening of strength development, or both; materials used to speed up the setting of mortar or concrete.

Accelerogram. The record from an accelerograph showing acceleration as a function of time.

Acceptance. (1) Manifestation that a party assents or agrees to a contract. (2) Approval of the work of a construction contract.

Access Control. A computerized security system designed to protect against unauthorized entry into buildings or building areas.

Access Door. A door or panel creating a means of access for the inspection or repair of concealed equipment.

Access Floor. A raised floor platform with removable panels to provide access to the area below.

Access Panel. Removable or swinging panel, usually flush with adjoining surface to provide access to concealed equipment or system components for inspection and maintenance purposes.

Access/Secure Control Unit. The controlling device of a computerized security system designed to protect against unauthorized entry into buildings or building areas.

Access Stair. A stair system to provide specific access to roofs, mechanical equipment rooms, or as a means of emergency exit.

Accessible. (1) As applied to equipment, admitting close approach because not guarded by locked doors (unless keys are readily available to those requiring access), elevation or other effective means. (2) As applied to wiring methods, not permanently closed in by the structure or finish of the building; capable of being removed without disturbing the building structure, finish, or fixed appurtenance thereto.

Accessible Heremetic. Assembly of motor and compressor inside a single bolted housing unit.

Accessible Location. A location which can be reached by standing on the floor, platform, runway, or other permanent working area.

Accessories (1) Tile accessories, ceramic or non-ceramic articles, affixed to or inserted in tile work, as exemplified by towel bars, paper, soap and tumbler holders, grab bars and the like. (2) Concrete accessories, implements or devices used in the formwork, pouring, spreading, and finishing, of concrete surfaces.

Accessory. An object or device aiding or contributing in a secondary way.

Accessory, Reinforcing. Items used to facilitate the installation of masonry or concrete reinforcing.

Accord and Satisfaction. Conduct of a debtor that indicates agreement to an amount of money owed by the debtor to a creditor.

Accordion Folding Door. A folding, hinged, or creased door with rollers which run along a track.

Accordion Partition. A folded, creased, or hinged interior dividing wall.

Account. A statement of transactions during a fiscal period and the resulting balance in each category of income and expense.

Account Balance. The difference between the sum of the debits and credits.

Accounting. The recording and auditing of financial accounts.

Accounting Period. The time that elapses between the preparation of financial statements.

Accounts Payable. Money owed by the firm to vendors for services or materials.

Accounts Receivable. Money owed to the firm for services rendered or for reimbursements.

Accoustical Plaster and Plastic. Sound absorbing finishing materials mill-formulated for application in areas where a reduction in sound reverberation or noise intensity is desired; these materials usually are applied to a minimum thickness of 1/2 inch and generally provide a noise reduction coefficient of at least .45 decibels.

Accretion. An increase of land area by the gradual or imperceptible action of natural forces.

Accrual Accounting. A method of keeping accounting records in which income is recorded when services are rendered and expenses are recorded when incurred, rather than when cash is received or paid out.

Accruals. The recognition of income and expenses as they occur even though they are not received or paid for until a later period.

Accrued Expenses. The entry into the liability accounts of expenses, incurred but not paid, at a given date.

Accrued Income. The entry into the asset accounts of income earned, but not received, at a given date.

Accrued Interest. Interest that has been earned but not yet paid.

Accrued Liabilities. Amounts representing liabilities incurred, but not paid, by a given date.

Accumulator. Storage tank which receives liquid refrigerant from evaporator and prevents it from flowing into suction line before vaporizing.

Acetic Acid. A compound, which in the pure state is a colorless, pungent, biting liquid; vinegar contains 4 to 12 percent of acetic acid.

Acetone. A water-white volatile, highly flammable solvent with ether-like odor; made by destructive distillation of hardwood, fermentation of butyl alcohol, and from petroleum sources; used extensively in making paint removers; dimethyl ketone; see Ketones.

Acetylene. A colorless hydrocarbon gas, burning with a bright flame, used as a fuel in welding and soldering.

Achromatic. (1) Having no color, being black, gray, or white. (2) Being neutral in color. (3) Difficult to color. (4) Lenses practically free from light of unwanted color.

Acid. A sour substance, one which liberates hydrogen ions in water and is sour and corrosive; will turn litmus red and has a pH of less than 7; acids are generally divided into two classes. (1) strong mineral or inorganic acids such as sulfamic, sulfuric, phosphoric, hydrochloric, or nitric, (2) weak organic or natural acids such as acetic (vinegar), citric (citrus fruit juices), oxalic, and fatty acids (oleic) such as palmitic and stearic.

Acid Condition In System. Condition in which refrigerant or oil in system is mixed with fluids that are acid in nature.

Acid Demand. Amount of acid required to lower pH and total alkalinity of pool water to correct level.

Acid Etch. (1) The use of acid to cut lines into metal or glass. (2) The use of acid to remove the surface of concrete.

Acidity. A general term applying to substances on the acid side of neutral.

Acid Number. A designation of the amount of free acid in oils, flats, waxes and resins, expressed as the number of milligrams of potassium hydroxide required to neutralize one gram of the material being tested.

Acid Rain. Sulfur dioxide emissions combining with water in the atmosphere and falling to the earth.

Acid Resisting Brick. Brick suitable for use in contact with chemicals, usually in conjunction with acid-resistant mortars.

Acid-Proof Counter. A horizontal work surface resistant to acid spills.

Acid-Resistant Grout. A grout that resists the effect of prolonged contact with acids.

Acid-Test Radio. A calculation of a firm's liquidity position; that is the ratio of its quick assets (readily convertible to cash) to current liabilities.

ACM. Asbestos-Containing Material.

ACORD 25-S. A form of insurance certificate issued by Agency Company Organization for Research and Development which has replaced the AIA standard form.

Acoustic. (1) Relating to sound or the sense of hearing. (2) The properties of a room or auditorium in transmitting sound.

Acoustical Block. A masonry block used for its sound-absorbing qualities.

Acoustical Materials. Those capable of absorbing sound waves.

Acoustical Panel. Ceiling and wall mounted modular units composed of sound absorbing materials.

Acoustical Tile. Ceiling panels in board form used for its sound absorbing properties, sometimes used on walls.

Acoustical Treatment. The act or process of applying acoustical materials to walls and ceilings.

Acoustical. Relating to sound or to the sense of hearing.

Acoustic Paint. Paint which absorbs or deadens sound.

Acoustics. The science of sound including its production, transmission, and effects.

Acre. (1) A piece of land measuring 43,560 square feet. (2) Unit for measuring land, equal to 43,560 square feet or 4840 square yards or 160 square rods.

Acre-Foot. A volume unit for measuring large quantities of water as in reservoirs and lakes; the amount of water that would cover one acre one foot deep, equal to 43,560 cubic feet.

Acre-Inch. One twelfth of an acre-foot.

ACRI. Air-Conditioning & Refrigeration Institute.

Across. The application of gypsum board where the long dimension is applied at right angles to the framing.

ACR Tubing. Tubing used in air conditioning and refrigeration; ends are sealed to keep tubing clean and dry.

Acrylic. (1) A general class of resinous polymers derived from esters, amides or other acrylic aid derivatives. (2) A transparent plastic material used in sheet form for window glass and skylights.

Acrylic Carpet. A carpet made from acrylic fiber composed of synthetic polymers.

Acrylic Plastics. Plastics based on resins made by the polymerization of acrylic monomers, such as ethyl acrylate and methecrylate.

Acrylic Resins. Family of synthetic resins made by polymerizing esters of acrylic acid; synthetic resins of excellent color and clarity used in both emulsion and solvent-based paints.

Acrylics. In carpeting, generic term including acrylic and modified acrylic (modacrylic) fibers; acrylic is a polymer composed of at least 85% by weight of acrylonitrile; modacrylic is a polymer composed of less than 85% but at least 35% by weight of acrylonitrile.

Acrylonitrile Butadiene-Styrene. Plastic material used in manufacturing drainage pipe and fittings.

Activated Carbon. Specially processed carbon used as a filter-drier; commonly used to clean air; pulverized carbon treated to be especially adsorbent.

Activated Charcoal. See Activated Carbon.

Activator. A catalyst, curing agent, or coreactant, as for an epoxy resin.

Active Door. In a pair of doors, the leaf that opens first and the one to which the lock is applied.

Active Pressure. The pressure exerted by retained earth; such as the earth retained by a retaining wall.

Active System. A solar heating or cooling system that requires outside mechanical power to move the collected heat.

Act of God. An unexpected event, not within the control of either party, that makes the performance of a contract impossible, unreasonable, or illegal.

Actual Dimension. The true size of a piece of lumber after it has been milled and dried; see Nominal Dimension.

Actual Notice. The giving of notice by mailing it or handing it to the recipient.

Actuator. That portion of a regulating valve which converts mechanical fluid, thermal energy, or electrical energy into mechanical motion to open or close the valve seats.

Acute Angle. An angle of less than 90 degrees.

Adapter. A mechanical device for connecting two different parts.

Adapter, Cubicle. See Cubical Adapter.

Adapter Terminal. Electrical fitting attached to the end of a conductor or to a piece of equipment, for taking power from an outlet in a way for which it was not designed.

Adaptive reuse. Adapting an old or historical building for a new purpose.

Addenda. A revision in the contract document made prior to the execution of the owner-contractor contract.

Additive. A substance added to another to impart different or special qualities; an admixture.

Additive Alternate. An alternate bid that, if accepted, adds to the contract price.

Adhered. Attached by adhesion, rather than mechanical anchorage, as adhered veneer.

Adhered Veneer. A veneer secured and supported through adhesion to an approved bonding material applied over an approved backing.

Adherence. The properties of bodies for sticking together.

Adherend. A body that is held to another body by an adhesive.

Adherend Failure. Failure of an adhesive joint when the separation is within the adherend.

Adhesion. 1. The state in which two surfaces are held together by interfacial forces which may consist of valence forces or interlocking action, or both; bonding strength; the attraction of a coating to the substrate, or of one coat of paint to another. 2. The soil quality of sticking to buckets, blades, and other parts of excavators.

Adhesion, Mechanical. Adhesion between surfaces in which the adhesive holds the parts together by interlocking action.

Adhesion Type Ceramic Veneer. Thinner sections of ceramic veneer, held in place by adhesion of mortar to unit and to backing; no metal anchors are required.

Adhesive. 1. A material capable of holding other materials together by surface attachment; glues, cements, pastes, epoxy, and mucilage are some of the common adhesives. 2. A compound, glue, or mastic used in the application of gypsum board products to framing or for laminating one or more layers of gypsum boards.

Adhesive Application. A means of applying gypsum board utilizing adhesives and supplemental mechanical fasteners.

Adhesive Bond. A relationship between two materials in contact with each other causing them to stick or adhere together by means other than cohesion.

Adhesive, Ceramic. Used for bonding tile to a surface; rubber solvents; rubber- and resin-based emulsions used as adhesives.

Adhesive Failure. Failure of an adhesive joint when the plane of separation is at the adhesive-adherend interface.

Adhesive, Pressure-Sensitive. An adhesive that will adhere to a surface at room temperature by briefly applied pressure alone.

Adhesive, Roof. A bonding agent used to cement roof materials.

Adhesive, Solvent. An adhesive having a volatile organic liquid as a vehicle, not including water-based adhesives.

Adhesive Spreader. A notched trowel used in the application of laminating adhesives.

Adhesive, Tile. Organic adhesive used for bonding tile to a surface; rubber solvents and resin-based and rubber emulsions can be used as adhesives.

Adhesive Wall Cups. Special clips or nails with large perforated bases for mastic application to most firm surfaces.

Adiabatic. 1.Impassable to heat. 2. Occurring without gain or loss of heat.

Adiabatic Compression. Compressing refrigerant gas without removing or adding heat.

Adiabatic Curing. The maintenance of ambient conditions during the setting and hardening of concrete so that heat is neither lost nor gained.

Adjacent. Nearby or adjoining.

Adjustable Bar Hanger. A metal hanger that can be made to fit the varying distances between floor and ceiling joists or rafters to securely hold electrical outlet boxes and devices.

Adjustable Shelf Standard. Metal items to support shelves usually in the form of strips attached to vertical surfaces.

Adjustable Shelf. A shelf that can be adjusted to different heights.

Adjustable Speed Motor. One in which the speed can be varied gradually over a considerable range, but when once adjusted remains practically unaffected by the load, such as shunt motors designed for a variation of field strength.

Adjustable Triangle. A transparent plastic drafting tool that can be adjusted and set for any angle.

Adjustable Wrench. An open faced wrench which can be adjusted to different sizes.

Admixture. 1. A material other than water, aggregates, and hydraulic cement used as an ingredient of concrete or mortar, and added immediately before or during its mixing. 2. A chemical additive used to alter the normal properties of concrete. 3. Any substance added to a plaster component or to plaster mortar for the purpose of altering its properties.

Adobe. Unburned or unfired brick, dried in the sun.

Adobe Masonry. Construction that utilizes unburned (unfired) clay masonry units.

AD Plywood. A designation or gradation of plywood. The A and the D designate quality of surface layers.

ADR. Alternative Dispute Resolution; includes mainly negotiation, mediation, and arbitration.

Adsorbed Water. Water held on surfaces in a material by either physical and/or chemical forces.

Adsorbent. Usually of a solid, having the ability to attract molecules of liquids, solutions, or gasses that adhere to its surface.

Adsorption. The process of attraction to a surface; the attachment of foreign molecules on the surface of a substance.

Adulteration. The addition of unwanted materials.

Ad Valorem. A tax imposed at a percentage rate of the value of the property, such as property tax.

Advancing Colors. Colors that give an illusion of being closer to the observer; warm colors in which red-orange predominates.

Adverse Possession. The overt occupation of real property under some claim of right that is opposed to the claim of some other claimant.

Advertisement for Bids. Published notice for receiving of bids for a construction project.

Advisor CM. A construction manager who is an advisor to the owner and who does not guarantee the construction cost.

Adze. A tool for cutting away the surface of wood, like an axe with an arched blade at right angles to the handle.

Aerate. To introduce air into a substance, such as into water at the kitchen sink.

Aeration. Act of combining substance with air.

Aerator. Device which adds air to water; fills flowing water with bubbles.

Aerembolism. Caisson Desease.

Aerial . 1. Relating to the air or atmosphere. 2. An antenna.

Aerobic. Activities or processes that can take place only in the presence of air or oxygen.

Aerosol. 1. A colloidal suspension of fine solid or liquid particles in gas, like smoke, fog, and mist. 2. A substance dispensed from a pressurized can in aerosol form.

Aesthetic Effect. Relating to the beautiful rather than to the merely pleasing, useful, or utilitarian; artistic and in accordance with the principles of good taste.

Aesthetics. Concerned with beauty, refinement, and good taste.

AEV. Automatic Expansion Valve.

Affidavit. A written statement that is made under oath.

A-Frame. A building structure where the main structural members forming the roof and floors are in an A-shape.

Aftershock. An earthquake occurring subsequent to a large earthquake, the main shock; the magnitude of an aftershock is usually smaller than the main shock.

Agent. One who acts with delegated authority for a principal.

Agglomeration. Formation of masses or aggregates of pigments; not dispersed.

Aggregate. 1. Inert particles such as sand, gravel, crushed stone, or expanded materials, in a concrete, plaster, or terrazzo mixture. 2. Granular material, such as sand, gravel, crushed stone, and iron blast-furnace slag, used with a cementing medium to form a hydraulic-cement, concrete or mortar. 3. Crushed stone, crushed slag, or water-worn gravel used for surfacing a built-up roof. 4. Any granular mineral material.

Aggregate Coated Panel. Sheet material, usually plywood, with decorative face of aggregate bonded with epoxy applied to one face.

Aggregate, Coarse. One of the four ingredients of concrete, usually gravel, which is retained on a #4 sieve.

Aggregate, Fine. One of the four ingredients of concrete, usually sand, which will pass the #4 sieve and will be retained on the #200 sieve.

Aggregate, Heavyweight. Aggregate of high specific gravity such as barite, magnetite, limonite, limenite, iron, or steel used to produce heavy concrete.

Aggregate, Lightweight. Aggregate of low specific gravity, such as expanded or sintered clay, shale, slate, diatomaceous shale, perlite, vermiculite, or slag; natural pumice, scoria, volcanic cinders, tuff, and diatomite, sintered fly ash, or industrial cinders; used to produce lightweight concrete; aggregate with a dry, loose weight of 70 pounds per cubic foot or less.

Aggregate Storage Bins. In a concrete batching plant, the bins that store the necessary aggregate sizes and feed them to the dryer in substantially the same proportions as are required in the finished mix.

Agitator. Device used to cause motion in confined fluid.

Agreement. A mutual understanding; a meeting of the minds; a contract duly executed and legally binding.

Agricultural Varnishes. Varnishes designed to protect and beautify farm implements and machinery.

AHU. Air Handling Unit.

AI. Asphalt Institute.

AIA. American Institute of Architects.

Air. 1. An invisible gaseous substance surrounding the earth, a mixture of mainly oxygen and nitrogen; the atmosphere. 2. An air conditioning system.

Air Adjusting Valve. Spray gun valve controlling input air.

Airborne Sound. Sound originating in a space; airborne sound can be created from the radiation of structure-borne sound into the air.

Air Break. In a plumbing system, a physical separation between a drain outlet from a fixture and an indirect waste receptor from the fixture.

Air Bubble. Bubble in paint film caused by entrapped air.

Air Cap. Perforated housing for atomizing air at head of spray gun; also called air nozzle.

Air Carbon-Arc Cutting. An arc-cutting process in which the severing of metals is effected by melting with the heat of an arc between an electrode and the base metal and an air stream is used to facilitate cutting.

Air Chamber. A short piece of pipe about 10 long, installed above the hot and cold valves of fixtures such as sinks, lavatories, and clothes washers which traps a column of air intended to cushion the rush of water as the valve is closed and prevents water hammer.

Air Change Method. A method of calculating the quantity of infiltration air into a building.

Air Changes per Hour. The number of times the air volume of a room or building can be replaced in an hour by an air-handling, circulating, or exhaust system.

Air Cleaner. Device used for removal of airborne impurities.

Air Cleaning. A control strategy to remove various airborne particulates and/or gases from the air; the three types of air cleaning most commonly used are particulate filtration, electrostatic precipitation, and gas absorption.

Air Coil. Coil on some types of heat pumps used either as an evaporator or a condenser.

Air Compressor. A mechanism which forces air at a high pressure into a storage tank where it is released through a regulator and a hose to power small tools.

Air Conditioner. Device used to control temperature, humidity, cleanliness, and movement of air in conditioned space.

Air Content. The amount of entrained or entrapped air in concrete or mortar, exclusive of pore space in aggregate particles, usually expressed as a percentage of total volume of concrete or mortar.

Air Cooled Compressor. The condenser component of a refrigeration system placed out of the refrigerant area in a series of copper tubes; a fan blows outdoor air across the tubes which contain the refrigerant.

Air-Cooled Condenser. Heat of compression is dissipated from condensing coils to surrounding air, by convection or by a fan or blower.

Air Cooler. Mechanism designed to lower temperature of air passing through it.

Air Core. Coil of wire not having a metal core.

Air Core Solenoid. Solenoid which has a hollow core instead of a solid core.

Air Diffuser. Air distribution outlet or grille designed to direct airflow into desired patterns.

Air Distribution. To force air to desired locations in a building or facility.

Air-Dried. Dried by exposure to air in a yard or shed without artificial heat.

Air-Dried Lumber. Lumber that has dried by being stored in yards or sheds for any length of time; for North America as a whole, the minimum moisture content of thoroughly air-dried lumber is 12 to 15%, and the average is higher.

Air Drying. Drying by oxidation or evaporating by simple exposure to air; used for drying block, brick, wood, or paint without any special equipment, simply by exposure to ambient air.

Air-Dry Weight. The unit weight of a light weight concrete specimen cured for seven days with neither loss nor gain of moisture at 60° F. to 80° F. and dried for 21 days in 50 plus or minus 7% relative humidity at 73.4° F. plus or minus 2° F

Air Eliminator. A mechanical device that expels excess air.

Air Embolism. Caisson Disease.

Air-Entertainment. The process by which air is introduced into a material while in a liquid or plastic state in the form of small isolated bubbles.

Air-Entrained Concrete. Concrete containing an admixture that produces microscopic air bubbles in the concrete; used to improve workability and freeze resistance.

Air Entraining Agent. A substance added to concrete, mortar or cement that produces air bubbles during mixing, making it easier to work with and increasing its resistance to frost and freezing.

Air Entraining Cement. A portland cement with an admixture that causes a controlled quantity of stable, microscopic air bubbles to form in the concrete during mixing.

Air Entrainment. Introduction of air into a process such as in concrete mixing or in a whirlpool bath.

Air Entrapment. Inclusion of air bubbles in paint film.

Air Exchange Rate. 1. The number of times that the outdoor air replaces the volume of air in a building per unit time, typically expressed as air changes per hour. 2. The number of times that the ventilation system replaces the air within a room or area within the building.

Airfield Marking. Lines, markers, or lines placed on airfield to aid in the takeoff and landing of planes.

Air Gap. 1. In a water supply system, the vertical distance from the top of the flood rim to the faucet or spout which supplies fresh water to the fixture; designed to prevent backsiphonage. 2. The space between magnetic poles or between rotating and stationary assemblies in a motor or generator.

Air Handler. Fan-blower, heat transfer coil, filter, and housing parts of a system.

Air Handling System. A system to heat, cool, humidify, dehumidify, filter, and transport air, consisting of an air handling unit, fresh air and exhaust air damper at the building exterior, ductwork, supply air, diffusers or registers, and return air grills in the conditioned space.

Air Handling Unit (AHU). Equipment that is designed to move conditioned air, containing fan(s), filter(s), heating coil(s), and/or cooling coil(s); units can be classified as either a central system or unitary; unitary equipment can be classified as rooftop unit, unitary package unit, unitary split system, or compound room unit.

Air Hose. Hose of air supply quality, usually red.

Air Infiltration. Leakage of air into rooms through cracks, windows, doors, and other openings.

Air Jet. In blast cleaning, type of blast cleaning gun in which the abrasive is conveyed to the gun by partial vacuum.

Airless Spraying. Spraying using hydraulic pressure to atomize the paint.

Air Lock. Air trapped within a pipe which restricts or blocks the flow of liquid through the pipe.

Air Manifold. Common air supply chamber for several lines.

Air Monitoring Test. A procedure used to determine the contents in a volume of air over a measurable period of time.

Air Nozzle. See Air Cap.

Air Plenum. Any space used to convey return air in a building or structure.

Air Powered Hoist. A hoist that is operated by compressed air.

Air Purger. A mechanical device that removes unwanted air.

Air Rights. The right to use the space above a piece of real property.

Air-Sensing Thermostat. Thermostat unit in which sensing element is located in refrigerated space.

Air-Slack. A condition where soft-body clay, after absorbing moisture and being exposed to the atmosphere, will spall a piece of clay and/or glaze.

Air Space. An open space or cavity in a wall or between building materials.

Air, Standard. See Standard Atmosphere.

Air Structure. A canvas structure supported by air produced by fans.

Air-To-Air Heat Exchanger. A method of heat recovery in which the intake and exhaust ducts are brought together at the heat exchanger; this system transfers only heat.

Air Tool. Attachments using compressed air to saw, spray-paint, sand, drill, or nail.

Air Transformer. Device for controlled reduction in air pressure.

Air Valve. Control valve in air line system.

Air Vent. 1. An opening in a building or structure for the passage of air. 2. Valve, either manual or automatic, to remove air from the highest point of a coil or piping assembly.

Air Voids. The small hollow spaces in cement paste caused by unwanted entrapped air bubbles and the smaller voids caused by air-entraining admixtures.

Air Volume. Quantity of air in cubic feet, usually per minute, at atmospheric pressure.

Air Washer. Device used to clean air while increasing or lowering its humidity.

Aisle. A passage space between rows of seats or tables; a passage space between rows of stored goods in a warehouse.

Alabaster. A compact translucent, usually white, form of gypsum, often carved into lamps, vases, and ornaments.

Alarm. A warning sound or device, as in smoke alarm or burglar alarm.

Alclad. Tradename for an aluminum alloy coated with pure aluminum to give high corrosion resistance.

Alcohol. A colorless volatile inflammable liquid, miscible with water, used as a solvent; in full, ethyl alcohol; the alcohols commonly used in painting are ethyl alcohol as a shellac solvent and methyl alcohol or wood alcohol in paint removers.

Alcohol Brine. Water and alcohol solution which remains a liquid below 32°F.

Alcohol Resisting. Showing no damage when in contact with alcohol.

Alcove. A small recessed section of a room or outdoor area.

Alfresco. In the open air.

Algae (plural). A non-flowering stemless water-plant; seaweed, pond scums, and phytoplankton; growing in water in the presence of sunlight and carbon dioxide.

Algaecide. Chemical that kills algae.

Algebra. 'The use of letters and other symbols to represent quantities in mathematical calculations.

Algistat. Chemical that inhibits algae growth.

Alienation. A conveyance of property to another.

Align. To be or come into precise adjustment or correct relative position.

Aliphatic. Describes a major class of organic compounds, many of which are useful as solvents.

Alizarin Lake. A bright red pigment with blue undertone made from the organic coal tar dyestuff, alizarin; some purple pigments are also marketed under this name.

Alkali. A chemical substance which effectively neutralizes acid material so as to form neutral salts; a base, the opposite of acid; examples are ammonia and caustic soda; highly destructive to oil paint films.

Alkali-Aggregate Reaction. The chemical reaction in concrete or mortar between alkalis (sodium and potassium oxides) from portland cement or other sources and certain constituents of some aggregates, primarily certain phases of silica, which causes deterioration in the form of strength loss, excessive expansion, and cracks in the concrete or mortar.

Alkaline. Having the properties of an alkali; having a pH of more than 7.

Alkalinity. Amount of bicarbonate, carbonate, or hydroxide compounds in water.

Alkali-Resistant Grout. A grout that resists the effect of prolonged contact with alkalis.

Alkyd. A synthetic resin, made usually with phthalic anhydride, glycerol and fatty acids from vegetable oils.

Allen Key. An Allen Wrench.

Allen Type Screw. Screw with recessed hex-shaped head.

Allen Wrench. An L-shaped hexagonal rod designed to fit into a hexagonal hole in a bolt head; also called an Allen Key.

Alley. 1. A narrow street or passageway behind buildings. 2. A building for bowling.

Alligatoring . 1. The cracking and crazing that occurs in asphalt roofing due to age and the effect of solar radiation and oxidation. 2. Coarse checking pattern characterized by a slipping of the new paint coating over the old coating to the extent that the old coating can be seen through the fissures, caused by improper build-up of paint. 3. In asphalt paving, i Interconnected cracks forming a series of small blocks resembling an alligator's skin or chicken-wire, caused by excessive deflection of the surface over unstable subgrade or lower courses of the pavement.

All-Inclusive Deed of Trust or Mortgage. A trust deed or mortgage that includes underlying financing; one payment is made to the all-inclusive mortgagee or beneficiary who then makes the payments on the underlying loans; also called a Wraparound Deed of Trust or Mortgage.

Allowable Bearing Capacity. The amount of pressure, expressed in pounds per square foot, that any particular soil will support, in the opinion of a soil mechanics engineer, taking into account acceptable settling and a safety factor.

Allowance. In the contract documents, a sum noted by the architect to be included in the contract sum, for a specific item; for example, a stipulated sum for carpeting or hardware which will be selected at a later date.

Allowances. The amount of tightness or looseness of male and female parts.

Alloy. 1. A substance composed of two or more metals. 2. Or of a metal and a nonmetallic constituent.

All-Purpose Compound. In gypsum wallboard installation, a joint treatment material that can be used as a bedding compound for tape, a finishing compound, and as a laminating adhesive or texturing product.

All Risk Policy. A property insurance policy that insures against all risks of loss that are not specifically excluded.

All Stretcher Bond. A brick bond showing only stretchers on the face of the wall with each stretcher divided evenly over the stretchers under it; staggered vertical joints.

Alluvium. Any geologically recent deposit of fine soil, silt, sand, and gravel from an ancient river.

Alpha Gypsum. A class of specially processed calcined gypsum having properties of low consistency and high strength.

Alternate. A stipulated construction item and its cost, in addition to the original base bid, for a specific item to be included or excluded from the project; a mutually agreed upon item that is used in place of the originally specified item.

Alternating Current (AC). Electric current in which direction of flow alternates or reverses in 60-cycle (Hertz) current, direction of flow reverses every 1/120th of a second.

Alternating Current Automatic Low Voltage Secondary Network. See Secondary Network.

Alternative Dispute Resolution (ADR). Resolution of a dispute without litigation, includes negotiation, mediation, and arbitration.

Alternator. An electric generator for producing alternating current.

Alumina. A mineral usually found in the clay used for brickmaking.

Alumina Cement . Used in high early strength concrete.

Alumina Porcelain. A vitreous ceramic whiteware for technical application in which Alumina ($A_{12}O_3$) is the essential crystalline phase.

Aluminous Cement. A hydraulic cement in which the principal constituents are calcium illuminates, instead of calcium silicates which comprise the major ingredients of portland cement. (See calcium aluminate cement)

Aluminum. A bluish silver-white malleable metallic element with good electrical and thermal conductivity, high reflectivity and resistance to oxidation.

Aluminum Extrusion. Aluminum sections formed by extrusion.

Aluminum Grid Walkway. Walkway fabricated of aluminum grid placed over roof surface to protect roofing surface from damage from traffic.

Aluminum Jacket. A watertight outer housing, fashioned from aluminum, placed around a pipe or vessel.

Aluminum Leaf. Aluminum hammered into very thin sheets.

Aluminum Oxide. 1. Corrosion on surface of an aluminum member caused by oxidation. 2. Hard and sharp abrasive made by fusing mineral Bauxite at high temperature.

Aluminum Paint. Mixture of finely divided aluminum particles in flake form combined with vehicle.

Aluminum Plate. Flat aluminum sheet material.

Aluminum Sheet Plate. Flat rolled aluminum plate.

Aluminum Silicate. White extender pigment made from China clay or feldspar, which provides very little color or opacity.

Aluminum Storefront. A facade of a building or structure which is constructed of a system of aluminum tubing and glass.

Aluminum Wire. Electrical conductors and cable manufactured from aluminum.

Aluminum, Zinc Coated. Aluminum, zinc plated for corrosion protection by hot dipping into molten zinc or by electrolysis.

Amalgamation. A merger or consolidation of two or more businesses.

Ambience. The feeling of the surroundings or atmosphere of a place.

Ambient Background Samples. Prevalent Level Samples.

Ambient Sound. The quiet-state noise level in a room or space, which is a composite of sounds from many external sources, both near and far, over which one individual has no control.

Ambient Temperature. 1. Room temperature or the temperature of the surroundings; temperature of fluid (usually air) which surrounds object on all sides.

Ambient. Surrounding on all sides; encompassing, as the ambient temperature.

Ambiguous. Having more than one meaning, as in a contract.

Ambo. Pulpit in a church.

Ambulatory. A sheltered place for walking, as in an aisle or cloister in a church or monastery.

Amenity. Pleasant or useful features in real estate that contribute satisfaction and pleasure rather than direct financial benefit.

American Bond. In masonry, a form of bond in which every sixth course is a header course and the intervening courses are stretcher courses.

American Gallon. A measure of liquid volume, 231 cubic inches.

American Lumber Standards. Rules for softwood lumber, dealing with recognized classifications, nomenclature, sizes, descriptions, amounts, shipping provisions, basic grades, grade marking, and inspection.

American Process Zinc Oxide. Zinc oxide pigment made directly from zinc ores; also called Direct Process.

American Standard Pipe Thread. Type of screw thread commonly used on pipe and fittings to assure a tight seal.

American Standards Association. Now known as American National Standards Institute.

American Vermillon. Chrome orange pigment.

Ames Taping Tools. Specially designed tools to mechanically apply taping compound and tape.

Amides. Curing agent combined with epoxy resins.

Amine Adduct. Amine curing agent combined with a portion of the resin.

Amines. Organic substituted ammonia; curing agent combined with epoxy resins.

Ammeter. An instrument for measuring electric current in amperes.

Ammonia. 1. Chemical combination of nitrogen and hydrogen, $NH_3$2. Popular refrigerant for industrial refrigerating systems; also a popular absorption system refrigerant; ammonia refrigerant is identified as R-117. 3. Ammonia combines with free chlorine in pool water to form chloramine which causes burning eyes, skin irritation and chlorine odor.

Amortization Schedule. A chart showing the monthly, quarterly, or annual payments necessary to pay off a loan including interest over a particular period of time.

Amortization. The gradual paying off of indebtedness by regular equal payments including interest and principal over a specified period of time.

Amp. Ampere.

Ampacity. Current-carrying capacity expressed in amperes.

Amperage. The strength of a current of electricity expressed in amperes; electron or current flow of one coulomb per second past given point in circuit.

Ampere. Unit of electric current equivalent to flow of one coulomb per second.

Ampere Turns. A measure of magnetic force; represents product of amperes times number of turns in coil of electromagnet.

Amplification. An increase in earthquake motion as a result of resonance of the natural period of vibration with that of the forcing vibration.

Amplifier. An electrical device to obtain amplification of voltage, current or power.

Amplitude. Maximum deviation from mean or center line of a wave.

Amyl Acetate. Banana oil; solvent for nitrocellulose, formed by esterification of acetic acid with amyl alcohol.

Anaerobic. An organism that can live without oxygen, such as the bacteria in a septic tank.

Anaglyph. A sculptured, chased, or embossed ornament worked in low relief.

Analog. A way of expressing one quantity in terms of another quantity; an analogous or parallel thing; for example, voltage, weight, or length to represent numbers, as watch hands represent time; compare with Digital.

Analogous Harmony. Colors which are related by containing one color in common; color harmony.

Analogy. Inference from a similar case.

Analysis . 1. Separation into constituent parts. 2. In engineering, the investigative determination of the detailed aspects of a particular phenomenon; may be qualitative, meaning a general evaluation of the nature of the phenomenon, or quantitative, meaning the numerical determination of the magnitude of the phenomenon.

Anchor. Irons or metals of special form and shapes used to fasten together and secure timbers or masonry.

Anchorage. 1. An attachment for resistance to movement; the movement can be a result of overturning, sliding or uplift; the most common anchorage for these movements are tie-downs (hold-downs) for overturning and uplift, and anchor bolts for sliding. 2. In posttensioning, a device used to anchor tendons to concrete member; in pretensioning, a device used to anchor tendons during hardening of concrete. 3. The securing of reinforcing bars in cast-in-place concrete either by hooks, bends, or embedment length.

Anchor Bolt. A bolt embedded in concrete for the purpose of fastening a building frame to a concrete or masonry foundation.

Anchor, Brick. Fasteners that are designed to attach and secure a veneer to a concrete or brick wall.

Anchor, Dovetail. A fastener with an interlocking joint that is wider at its end than at its base.

Anchored Type Ceramic Veneer. Thicker sections of ceramic veneer held in place by grout and wire anchors connected to backing wall.

Anchoring Cement. Grout used in sleeves to anchor tubing in place.

Anchor Pattern. Profile surface roughness usually attained by blasting.

Anchor, Rafter. A bolt or fastening device which attaches the rafters to the walls or rafter plate.

Anchor Slot. A groove in an object into which a fastener or connector is inserted to attach objects together.

Anchor Tenant. The main tenant in a shopping center development; a large stable tenant or tenants that are expected to attract other tenants and customers to the development.

Anchor Ties. Any type of fastener used to secure wood framing, steel, or masonry to some stable object, such as a foundation or another wall; usually for tension value.

Anchor, Wood. A bolt or fastening device which attaches wood to wood or wood to other materials.

Andalusite. A polymorph, along with sellimanite and kyanite, of composition $A_{12}O_3SiO_2$; in firing, it dissociates to yield principally mullite.

Andiron. A metal stand, usually a pair, for supporting wood fuel in a fireplace.

Anechoic Chamber. A room or building that is free from echoes and reverberations.

Anemometer. Instrument for measuring the rate of air flow or motion.

Aneroid Barometer. An instrument for measuring atmospheric air pressure by its action on the elastic lid of an evacuated box. This is the basis of a surveyor's aneroid barometer which is used to measure altitude; an altimeter.

Anesthetizing Location. Areas in hospitals in which flammable anesthetics are or may be administered to patients; such locations include operating rooms, delivery rooms, and anesthesia rooms, and will also include any corridors, utility rooms or other areas which are or may be used for administering flammable anesthetics to patients.

Angle. 1. The difference in direction of two lines which meet or tend to meet, usually measured in degrees. 2. An inside corner. 3. A structural section of steel which resembles an L in cross section. 4. In masonry, a portion of a whole brick which is used to close the bond or brickwork at corners.

Angle Bar. A steel structural member in the shape of an L; classified by the thickness of the stock and the length of the legs.

Angle Blasting. Blast cleaning at angles less than 90 degrees.

Angle Block. A square of tile specially made for changing direction of the trim.

Angle Brick. Any brick shaped to an oblique angle to fit a salient corner.

Angle Divider. A toolused by tilesetters to determine the degree of an angle to cut; used for fitting trim, moldings, and floors into corners; a corner angle is measured by adjusting the divider to fit the corner.

Angle Float. A concrete or plastering finishing tool having a surface bent to form a right angle; used to finish re-entrant angles.

Angle Iron. A rolled steel structural member with an L-shaped section; used to support brickwork over doors and windows and is sometimes used as main runners in lieu of channels to support plaster.

Angle of Degree. On an airless spray cap, the orifice angle; controls width of spray and pattern angle.

Angle of Incidence. The angle that a line or light ray striking a surface makes with the perpendicular at that point, the point of incidence.

Angle of Reflection. The angle that a reflected ray makes with the perpendicular to a surface at the point of incidence.

Angle of Repose. 1. The natural angle that a pile of a material, such as earth, sand, or gravel will assume. 2. The angle at which a body will slide down an inclined plane impelled only by gravity.

Angle, Shelf. A structural angle which is fastened to the structure of a building to support a wall or other component.

Angle Steel. An L-shaped steel member.

Angle Stop. A water valve that occurs where there is a 90 degree change in direction of the piping, as under a lavatory or sink.

Anhydrite. The mineral consisting primarily of anhydrous calcium sulfate, $CaSO_4$

Anhydrous. Dry; free from moisture or water in any form.

Anhydrous Calcined Gypsum. Keene's Cement.

Anhydrous Calcium Sulphate. A stable form of gypsum from which practically all of the water of crystallization has been removed; also called dead-burned gypsum; dry chemical made of calcium, sulphur and oxygen, $CaSO_4$.

Aniline Colors. Coal-tar derivatives precipitated on a colorless base.

Anisotropic. Exhibiting different properties when measured along different axes; in general, fibrous materials such as wood are anisotropic; not isotropic.

Annealed. Cooled under controlled conditions to minimize internal stresses.

Annealed Glass. Glass whose surface has been heated to melting and then cooled to provide a toughened surface; also called heat strengthened glass.

Annealing, Process of heat treating metal to get desired properties of softness and ductility.

Annex. A wing or an addition to a building.

Annual Growth Ring. The layer of wood growth put on a tree during a single growing season, comprised of springwood and summerwood.

Annuity. An investment of money entitling the investor to receive a series of equal monthly or yearly payments.

Annular Ring Nail. A nail that has grooves around the shank which prevent easy withdrawal; also called a ring shank nail.

Annunciator Panel. A panel mounted on a surface which indicates by lights which circuits have been activated.

Annunciator. A electromagnetic device used to show which of several circuits are activated.

Anode. 1. The positive electrode in an electrolytic cell. 2. The negative terminal of a primary cell such as a battery.

Anodize. Coat a metal, such as aluminum, with a protective oxide layer by electrolysis.

Anodized. A metal that has been subjected to electrolytic action in order to coat with a protective or decorative film.

Anodized Door. A door which has been given an aluminum oxide coating by electrolytic action.

Anodizing. An electrolytic process that forms a permanent, protective coating on aluminum.

Anodized Tile. Tile which has been given an aluminum oxide coating by electrolytic action.

Anomaly. An irregularity, abnormality, peculiarity, or deviation from the norm.

ANSI. American National Standards Institute, Inc.

Antenna. A metallic device used for radiating or receiving radio waves.

Anterior. Situated before or toward the front.

Anteroom. A small room leading to a main room; a waiting room.

Anticipatory Breach. A positive statement by a party to a contract that the party will not perform the terms of the contract.

Anticlastic. Saddle-shaped, or having curvature in two opposing directions.

Anti-Corrosive Paint. Metal paint designed to inhibit corrosion; applied directly to the metal.

Anti-Flooding Agent. A synthetic organic product used to reduce floating and flooding of iron blues, carbon blacks, and chrome greens.

Anti-Fouling Paint. A special coating for ship bottoms, containing poison like copper or mercury, formulated to effect the release of the poison at a controlled rate, to prevent attachment and growth of marine organisms such as barnacles and algae.

Anti-Friction Bearing. Any bearing having the capacity of effectively reducing friction.

Anti-Friction Latch Bolt. A latch bolt designed to reduce friction when the bolt starts to engage the lock strike.

Antihammer Device. An air chamber such as a closed length of pipe or a coil which is designed to absorb the shock caused by closing a valve rapidly.

Antimicrobial. Agent that kills microbial growth.

Antimony Oxide. Pure white pigment which provides about same hiding power as lithopone.

Antioxidant . Protective compound used on ends of aluminum wiring at connections to prevent corrosion that would interfere with a solid safe connection.

Antiquing. Furniture finishing technique intended to give appearance of age or wear.

Anti-Siphon Trap. A trap which is designed to prevent the siphonage of its water seal by increasing the diameter of the outlet leg of the trap so that it contains a sufficient volume of water to prevent a siphoning action.

Anti-Skinning Agent. A synthetic organic product, used to prevent forming of surface skin in packaged varnishes and paints.

Apartment. A suite of rooms, usually let as a dwelling; a dwelling unit.

Apartment Hotel. A hotel containing apartments as well as transient rooms.

Apartment House. A building containing three or more separate residential apartments; also called an Apartment Building.

Aperture. 1. Any opening. 2. An opening that varies the amount of light entering a camera.

Apex. The highest point; vertex.

Apices. Plural of apex.

Appliance. An appliance is current-consuming utilization equipment, generally other than industrial, normally built in standardized sizes or types, which is installed or connected as a unit to perform one or more functions such as clothes washing, air conditioning, food mixing, or deep frying.

Appliance, Fixed. An appliance which is fastened or otherwise secured at a specific location.

Appliance, Portable. An appliance which is movable or can easily be moved from one place to another in normal use.

Appliance, Stationary. An appliance which is not easily moved from one place to another in normal use.

Application for Payment. A written document forwarded by the contractor requesting payment for work completed.

Application Rate. The quantity (mass, volume or thickness) of material applied per unit area.

Applicator. 1. One who applies. 2. Tool for applying.

Applied Preservative Treated. Applied treatment of wood or plywood to make it resistant to deterioration from moisture and insects.

Appraisal. A valuation of property by a qualified person, usually in the form of a written report.

Appraise. To give an expert judgment of the value or merit of.

Appreciation. The increase in value of a property from any cause.

Apprentice. A person who is learning a trade, art or calling by being employed in it for an agreed period for low wages.

Approved. Acceptable to that enforcing agency having the responsibility and authority to grant the approval for the specified condition or application.

APR. Annual Percentage Rate.

Apron. 1. The lower trim member under the sill of the interior casing of a window. 2. An upward or downward vertical extension of a sink or lavatory. 3. A paved area immediately adjacent to a building, structure or facility.

Apron Wall. That part of a panel wall between the window sill and the support of the panel wall.

Apse. A projecting part of a building, such as in a church, usually semicircular and vaulted.

APWA. American Public Works Association.

Aqueduct. 1. An artificial channel or conduit for conveying large quantities of water, usually supported by a bridge-like structure. 2. A structure for conveying a canal over a river or a valley.

Aqueous. Containing water.

Aquifer. A strata of porous permeable rock or soil that is capable of holding a large quantity of water.

AR. Acid Resisting.

Arabesque. An ornament, style or design of intertwined leaves, flowers, or fruit to produce an intricate pattern of interlaced lines.

Arabic Numeral. One of the number symbols, 0, 1, 2, 3, 4, 5, 6, 7, 8, 9.

Arbitration. 1. A proceeding for resolution of disputes in which a neutral person or panel, after hearing evidence presented by both sides, makes a final and binding decision that resolves the dispute. 2. A hearing used to resolve disputes.

Arbitration Award. The arbitrator's judgment or decision in an arbitration.

Arbitration Demand. A written notice by one party to a contract served on the other to avail of the arbitration provisions of the contract.

Arbor. 1. A light open-work structure of wood or metal, covered or intended to be covered with vines, as in a park or garden. 2. An Axle or spindle on which something revolves. 3. A device holding a tool in a lathe.

Arc. 1. Arched or curved. 2. A portion of an ellipse or circle.

Arc Blow. The deflection of an electric arc from its normal path because of magnetic forces.

Arc Voltage. The voltage (electrical potential) across the welding arc.

Arc Welding. A process of joining two pieces of metal by melting them together at their interface with a continuous electric spark and adding a controlled additional amount of molten metal from a metallic electrode.

Arcade. 1. An arched covered passageway. 2. An amusement center with coin-operated games.

Arch. A curved structural member used to span an opening or recess; also built flat. Structurally, an arch is a piece or assemblage of pieces so arranged over an opening that the supported load is resolved into pressures on the side supports, and practically normal to their faces.

Arch Brick. Wedge-shaped brick for special use in an arch.

Arch Corner Bead. A job-shaped length of corner bead used to define the curved portion of arched openings.

Arch Culvert. A curved shaped drain under a roadway, canal or embankment.

Archeology. The study of human history and prehistory through the excavation of sites and the analysis of human remains.

Archetype. An original model; a Prototype.

Archimedes' Principle. A law of physics that states that a floating body displaces a weight of liquid equal to its own weight.

Archimedes' Screw. A primitive device for raising water, consisting of a pipe spirally wound around an axis.

Archipelago. A group of islands in the sea.

Architect. One who designs buildings and advises on their construction.

Architectonic. 1. Architectural. 2. Having an organized and unified structure or concept that suggests an architectural design.

Architectonics. The science of architecture.

Architect's Project Representative. A more or less continuous architect's representative on the jobsite; reports to the architect; formerly called Clerk of the Works.

Architectural. 1. Of or relating to or conforming to the rules of architecture. 2. Having a single, unified overall design, form, or structure.

Architectural Equipment. The implements, apparatus, or equipment used in the construction and initial outfitting of a building.

Architectural Fee. The amount of money charged by an architect for professional services such as programming, design, preparation of contract documents, and administration of the construction of a building or facility.

Architectural Terra Cota. Hard-burned, glazed or unglazed clay building units, plain or ornamental, machine-extruded or hand-molded, and generally larger in size than brick or facing tile; also see Ceramic Veneer.

Architectural Woodwork. Finish work using wood or composites for ornamental designs or casework.

Architrave. 1. The lowest division of an entablature. 2. The molding around a rectangular opening, such as a door or window.

Are. A metric unit of measure equal to 100 square meters. 100 ares equal 1 hectare.

Area. 1. The extent or measure of a surface, expressed in square units. 2. The space allocated to a particular function.

Area Divider. In roof construction, a raised double wood member attached to a properly flashed wood base plate that is anchored to the roof deck; used to relieve the stresses of thermal expansion and contraction in a roof system where no expansion joints have been provided.

Area Drain. Any drain installed for the purpose of collecting rain water from an open area and channeling it to the storm drain.

Area Wall. 1. The masonry surrounding or partly surrounding an area. 2. The retaining wall around basement windows below grade.

Areaway. A sunken space providing access, air, and light to a subterranean building area.

Ark. A repository in a synagogue for the scrolls of the Torah.

Armature. Part of an electric motor, generator, or other device moved by magnetism.

Armored Cable. Metal sheathed flexible electrical cable; BX cable.

Armored Faceplate. A metal plate which is fastened into the strike at the door jamb to provide a protection for the lock and keeper mechanisms.

Armor Plate. A kick-plate made of metal installed on the bottom of a door to protect it from denting and scratching.

Aromatic. Derived from or belonging to a major class of organic compounds, many of which are useful as solvents.

Aromatic Hydrocarbons. Strong solvents such as benzene, toluene, xylene.

Arrester, Lightning. A device connected to an electrical system to protect from lightning and/or voltage surges.

Arris. The sharp edge or salient angle formed by the meeting of two surfaces, as in a molding.

Art. Work exhibiting human creative skill or its application; creative activity; human skill or workmanship; an occupation requiring knowledge or skill.

Arterial Streets. Primary surface roads connecting to expressways with on and off ramps, also connecting sectors of cities, with surface crossings (controlled by traffic lights), normally over continuous long stretches of the cityscape, restricted for parking and to direct access often to adjacent commercial developments; pedestrian crossing controlled.

Artesian Well. 1.A bored well from which water flows from internal pressure. 2. A deep- bored well.

Artificer. 1. A skilled or artistic craft worker. 2. An inventor.

Artificial Intelligence. The ability of a computer to think and work like a human being; at present no computer has full artificial intelligence.

Artificial or Accelerated Drying. The process of drying block or brick with relatively warm, dry air, or other means.

Artificial Pozzolan. Fly ash and other similar substances such as rice hull ash and microsilica.

Artisan. A skilled worker in a trade such as carpentry, plumbing, or painting.

Asbestine. Natural fibrous magnesium silica, which is pure white in color; used as an extender pigment in paints.

Asbestos. A group of natural, fibrous, impure silicate materials, formerly used for its noncombustible, non-conducting, or chemically resistant properties; no longer used as it causes Asbestosis

Asbestos Abatement. Procedures to control fiber release from asbestos-containing materials in a building or to remove it entirely; may involve removal, encapsulation, repair, enclosure, encasement, and operations and maintenance programs.

Asbestos Cement. Cement asbestos.

Asbestos-Containing Material (ACM). Any material containing more than one percent asbestos.

Asbestos-Containing Material. Any material containing more than one percent asbestos.

Asbestos-Free Compound. Joint treatment products that have no asbestos fiber.

Asbestosis. A lung disease caused by inhaling asbestos particles.

Asbestos Program Manager. A building owner or designated representative who supervises all aspects of the facility asbestos management and control program.

Asbestos Removal. A special trade or occupation for the removal and disposal of hazardous asbestos.

Asbestos Vacuum. A filtered device using air suction to collect wetted down asbestos for easy removal into specially made bags.

As-Built Drawings. See Record Drawings.

As-Builts. See Record Drawings.

Aseismic Region. A region that is relatively free of earthquakes.

Ash Dump. A metal access door in the floor of a fireplace firebox to dump the ashes into the ash pit below.

Ashlar. A pattern of masonry consisting of squared stones.

Ashlar Veneer. An ornamental or protective facing of masonry composed of squared stones.

Ash Pit. The compartment below a fireplace in which ashes may be dumped from the firebox and stored for later removal.

ASHRAE. American Society of Heating, Refrigeration, and Air-Conditioning Engineers, Inc.

Askarael. A synthetic nonflammable insulating liquid which, when decomposed by the electric arc, evolves only nonflammable gaseous mixtures.

ASME Boiler Code. Standard specifications issued by the American Society of Mechanical Engineers for the construction of boilers.

Aspect Ratio. Ratio of length to width of a rectangular air grille or duct.

Asphalt. A brown to black bituminous pitch occurring naturally or made from a residue of petroleum refining; also called asphaltum; used for pavements, roofing, and as a waterproofing cement; asphalt comes in a wide range of viscosities and softening points, from about 135° F (used on dead level roofs) to 210° F. or more (to be used on special steep roofs); asphalt is a constituent in varying proportions of most crude petroleums; see Bitumen.

Asphalt Cement. A fluxed or unfluxed asphalt specially prepared as to quality and consistency for direct use in the manufacture of asphalt pavements.

Asphalt Curb. An extruded berm made from asphaltic concrete.

Asphalt Cut Back. Asphalt plus thinner; asphalt solution; asphalt coating formed by dissolving asphalt.

Asphalt Dampproofing. The application of asphalt to act as a water-resisting treatment to the surface of a concrete or masonry wall.

Asphalt Demolition. The destruction of roadways constructed of asphalt using large wheeled machinery that tears up the old asphalt into manageable pieces for easy removal.

Asphalt Emulsion. Asphalt dispersion; not a solution; a water emulsion of asphalt.

Asphalt Expansion Joint. Felt or fiberboard premolded and impregnated with asphalt and used as an expansion joint for cast-in-place concrete.

Asphaltenes. The high molecular weight hydrocarbon fraction precipitated from asphalt by a designated paraffinic naphtha solvent at a specified solvent-asphalt ratio.

Asphalt Felt. An asphalt-saturated felt or an asphalt-coated felt.

Asphlatic Concrete. High quality, thoroughly controlled hot mixture of asphalt cement and well-graded, high quality aggregate, thoroughly compacted into a uniform dense mass.

Asphaltic Paint. 1.A liquid form of asphalt which can contain other materials such as aluminum flakes and mineral pigments, generally used for creating a water-resistant coating. 2. A liquid product used for weatherproofing.

Asphalt Leveling Course. A course asphalt aggregate mixture of variable thickness used to eliminate irregularities in the contour of an existing surface prior to a superimposed treatment or construction.

Asphalt Mastic. A mixture of asphaltic material, mineral aggregates, and fine mineral aggregates that can be poured when heated but needs mechanical manipulation to apply when cool.

Asphalt Membrane. A layer of asphalt used on a flat roof.

Asphalt Prime Coat. A low viscosity liquid asphalt applied to prepare an untreated surface to penetrate into the voids, harden the top and help bind it to the overlying asphalt surface.

Asphalt Repair. The act or process of patching worn road surfaces or leaking roofs.

Asphalt Rock. Porous rock such as sandstone or limestone that has become impregnated with natural asphalt through geologic process; also called Rock Asphalt.

Asphalt Roof. A roof system which uses asphalt materials as a covering.

Asphalt Saturated Felt. (1) A building and roofing felt sheet impregnated with a bituminous waterproofing material. (2) A moisture-resistant sheet material, available in several different thicknesses, usually consisting of a heavy paper that has been impregnated with asphalt.

Asphalt Shingle. Saturated roofing felt either in large rolls or cut into composition shingles, impregnated with aggregate particles applied to a roof surfaces.

Asphalt Subgrade, Improved. Subgrade, improved as a working platform (1) by the incorporation of granular materials or stabilizers such as asphalt, lime, or portland cement, prepared to support a structure or a pavement system, or (2) any course or courses of select or improved material placed on the subgrade soil below the pavement structure; subgrade improvement does not affect the design thickness of the pavement structure.

Asphalt Surface. A flat layer of asphalt.

Aspirating Psychrometer. Device which draws sample of air through it to measure humidity.

Assemblage. 1. Something put together from parts. 2. A random, unordered assemblage is called a gathering. 3. An ordered assemblage is a system.

Assembly Area. An outdoor place for the gathering of a group of people exiting from a building or for an educational, sporting, or entertainment event.

Assembly Room. A room for the gathering of a group of people, such as an auditorium, gymnasium, restaurant, or meeting room.

Assessed Value. The value of a parcel of real property, usually a fraction of the market value, used for the purpose of determining the amount of property tax.

Asset. A possession or resource having value.

Assignment. Transfer of the rights and duties under a contract from one party to another.

ASTM Standards. Materials specifications issued by the American Society of Testing Materials.

Astragal Weatherstripping. Fabric, rubber or plastic strips attached to the molding that is attached to one of a pair of doors or casement windows to cover the joint between the two stiles.

Astragal. An interior molding attached to one of a pair of doors or window sash in order to prevent swinging through; also used with sliding doors to insure tighter fitting where doors meet.

Asymmetrical. Not symmetrical; asymmetry.

Athletic Equipment. Various devices used for exercising or the playing of sporting events.

Atmospheric Corrosion. Galvanic corrosion that occurs between two adjoining dissimilar metals in a humid atmospheric condition.

Atmospheric Pressure. The pressure that atmospheric gases in air exert upon the earth; measured in pounds per square inch or grams per square centimeter;one standard atmosphere equals 14.69 psi of pressure and measures 760 mm (29.92 inches) in a barometer of mercury.

Atmospheric Vacuum Breaker. A simple mechanical device consisting essentially of a check valve in the supply line and a valve member (on the discharge side of the check valve) opening to the atmosphere when the pressure in the line drops to atmospheric; also called a siphon breaker.

Atom. The smallest particle of a chemical element; considered as a source of vast potential energy.

Atomic-Hydrogen Welding. An arc-welding process wherein coalescence is produced by heating with an electric arc maintained between two metal electrodes in an atmosphere of hydrogen; shielding is obtained from the hydrogen; pressure may or may not be used and filler metal may or may not be applied.

Atomize. Process of changing a liquid to minute particles or a fine spray; break steam into small particles.

Atomizer. Device by which air is introduced into material at the nozzle to regulate the texture of machine-applied plaster.

Atrium. A central courtyard with surrounding rooms opening off it.

Attachment Bug. A device which, by insertion in a receptacle, establishes connection between the conductors of the attached flexible cord and the conductors connected permanently to the receptacle.

Attenuation. The reduction of the energy or intensity of sound.

Attenuation Blanket. Material utilized to help in the reduction of the energy or intensity of sound.

Attic. A room or space immediately below the roof of a building.

Attic Draft Stop. A partition in an attic, from roof to ceiling level, dividing the attic into discrete areas to prevent the spread of smoke and drafts.

Attic Insulation. Treated shreds of cellulose material that are blown into attic spaces or fiberglass rolls that are rolled out between ceiling joists to aid in weatherproofing a building or facility.

Attic Ventilators. 1. Openings in the roof or in gables for the purpose of allowing air to circulate. 2. Mechanical devices with power-driven fans to force the circulation of air.

Attorney. A qualified lawyer licensed to practice.

Attorney-in-Fact. A person acting for another under the authority of a power of attorney.

Audio. Acoustical, mechanical or electrical frequencies used in the transmission, reception or reproduction of sound.

Audio Cable. A cable over which the transmission, reception, or reproduction of sound is carried.

Audio Sensor. A device that responds to the physical stimulus of sound and transmits a resulting impulse.

Audio Visual. Involving both sight and sound.

Audit. A formal examination and report of a person's or entity's financial condition, usually by a qualified certified public accountant.

Auditorium. A building or room where an audience sits.

Auger. An instrument or device used for boring or forcing through materials or soil.

Auger, Fence. A rotating drill with a screw thread used to drill deep, straight, and narrow holes for the installation of fence posts.

Authorized Person. One who is properly authorized to perform specific duties under the conditions existing, usually in relation to a contract.

Autoclave. A pressure vessel in which an environment of superheated steam at high pressure may be produced; used for sterilizing, assisting a chemical reaction, in the curing of concrete products, and in the testing of hydraulic cement.

Autoclaved Aerated Concrete. A factory-produced building stone; precast, lightweight concrete, in the form of large building blocks, panels, or planks; made of closed cell concrete steam cured in a pressurized autoclave.

Autoclave Curing. Steam curing of concrete products, sand-lime brick, asbestos-cement products, hydrous calcium silicate insulation products, or cement in an autoclave at maximum ambient temperatures generally between 340 and 420° F (170-215° C); also called High Pressure Steam Curing.

Auto Lift. 1. An apparatus for lifting automobiles in order to have access to the carriage underneath. 2. An apparatus used to move automobiles up and down in a parking structure.

Auto Transfer Switch. An electrical transfer switch that operates automatically.

Automatic. 1. A device that works by itself without human intervention, usually actuated by some impersonal influence, as for example, a change in current strength, pressure, temperature, or mechanical configuration. 2. Done spontaneously without conscious intention, see Non-Automatic.

Automatic Control. Valve action reached through self-operated or self-actuated means, not requiring manual adjustment.

Automatic Cyclic Control. A control system in which the opening and closing of the weigh hopper discharge gate, the bituminous discharge valve, and the pugmill discharge gate are actuated by means of self-acting mechanical or electrical machinery without any intermediate manual control. The system includes preset timing devices to control the desired periods of dry and wet mixing cycles.

Automatic Defrost. System of removing ice and frost from evaporators automatically.

Automatic Door. A door equipped with a power-operated mechanism and controls that open and close the door automatically upon receipt of a momentary actuating signal; the switch that begins the automatic cycle may be a photoelectric device, floor mat, or manual switch.

Automatic Dryer Control. In a concrete batching plant, a system that automatically maintains the temperature of aggregates discharged from the dryer within a preset range.

Automatic Expansion Valve (AEV). Pressure-controlled valve which reduces high-pressure liquid refrigerant to low-pressure liquid refrigerant.

Automatic Frost Control. Control which automatically cycles refrigerating system to remove frost formation on evaporator.

Automatic Ice Cube Maker. Refrigerating mechanism designed to automatically produce ice cubes in quantity.

Automatic Proportioning Control. In an asphaltic concrete plant, a system in which proportions of the aggregate and asphalt fractions are controlled by means of gates or valves which are opened and closed by means of self- acting mechanical or electronic machinery without any intermediate manual control.

Automatic Skimmer. See Surface Skimmer.

Automation. The use of automatic equipment to save mental and manual labor.

Automaton. A piece of mechanism with concealed motive power; a Robot.

Automotive Equipment. Implements or apparatus that pertain to the repair and care of the automobile.

Automotive Hoist. 1. An apparatus for hoisting automobiles in order to have access to the carriage underneath. 2. An elevator used to move automobiles up and down in a parking structure.

Autopsy Table. Surface upon which a body is placed for examination after death; a specified item installed in hospitals, laboratories, and police crime labs.

Autotransformer. Transformer in which both primary and secondary coils have turns in common; step-up or step-down of voltage is accomplished by taps on common winding.

Auxiliary Heat. The extra heat provided by a conventional heating system for periods of cloudiness or intense cold, when a solar heating system is insufficient.

Auxiliary Members. In a concrete shell structure, ribs or edge beams which serve to strengthen, stiffen or support the shell; usually, auxiliary members act jointly with the shell.

Auxiliary Switch. A standby device for switching.

Available Chlorine. Free or combined chlorine used to disinfect pool water.

Avenue. An access road or street.

Average. 1. An arithmetic mean. 2. An amount obtained by adding the elements of a set and dividing by the number in the set. 3. The ordinary standard.

Aviary. An enclosure or building for keeping birds confined.

Avoirdupois. 1. Weight. 2. Avoirdupois Weight.

Avoirdupois Weight. The weight system based on the pound of 16 ounces and the ounce of 16 drams.

Award. 1. Arbitration award. 2. Contract award.

Awarding Authority. The owner or the agent of the owner who awards an original building or construction contract, also known as the prime contract; this term is generally used with public works.

AWG. American Wire Gauge.

AWI. Architectural Woodwork Institute.

Awl. A small sharp pointed hand tool for marking lines or piercing small holes.

Awning Window. A window hinged at the top.

Awning. A sheet of canvas, plastic, or metal stretched on a frame to give shelter over a window, doorway, or other area.

Axial. In a direction parallel to the long axis of a structural member; in line with the axis; an axial force causes tension or compression stress over the cross section of a member.

Axial Load. Force directly coincident with the primary axis of a structural member such as a beam.

Axial Stresses. Compressive or tensile stresses that are coincidental with the central axis of a structural member.

Axis. A straight line of reference; in three dimensions, the three axes are referred to as x, y, and z.

Axis, Neutral. Centroidal axis, transverse to longitudinal axis of a structural member, which is neither stretched nor shortened by bending of the member.

Axis of a Weld. An imaginary line through the length of a weld perpendicular to the cross section at its center of gravity.

Axis of Symmetry. A line dividing an area into two similar but opposite handed figures.

Axle. A fixed or rotating rod or spindle on which a wheel or group of wheels rotate.

Axminster Carpet. One of the four basic weaves used in making carpets; woven on an Axminster loom, the pile tufts in this weave are mechanically inserted and bound to the back in a manner similar to the hand knotting of Oriental rugs, making possible almost unlimited combinations of colors and patterns; see Woven Carpet.

Axonometric Projection. A drafting projection in which objects on the drawing appear inclined with three sides showing and with horizontal and vertical distances drawn to scale but diagonal and curved lines distorted.

Azeotrope. Having constant maximum and minimum boiling points.

Azimuth. The horizontal angle or direction of a compass bearing.

B

Back. 1. The part or surface behind the front; the rear. 2. See Steel Square. 3. The back surface of gypsum board; the opposite side from the face; usually the side that would be concealed or the side in contact with the framing; also referred to as the Code Side.

Backband. A simple molding sometimes used around the outer edge of plain rectangular casing as a decorative feature.

Back Bar. A counter, shelf, or shelving for bottles and glasses extending along the wall of a barroom behind the bar.

Back Blocking. A short piece of gypsum wallboard adhesively laminated behind the joints between each framing member to reinforce the joint.

Backboard. A vertical structure used as a guard against losing thrown objects, usually specified in the design of athletic playing fields and surfaces.

Backcharge. An offsetting charge against a bill, often asserted by an owner against a prime contractor or a by prime contractor against a subcontractor based on supposedly defective construction work.

Backcheck. A valve that allows flow of a liquid in only one direction; a backflow preventer; a backcheck valve.

Back Clip. Specially designed clips attached to the back of gypsum board that fit into slots or other formations in the framing to hold the gypsum board in place; often used in demountable partition designs.

Backer Board. See Backing Board.

Backfill. Earth or earthen material used to fill the excavation around a foundation; the act of filling around a foundation.

Back Filling. 1. Rough masonry built behind a facing or between two faces. 2. Filling over the extrados of an arch. 3. Masonry in spaces between structural timbers, sometimes called Nogging. 4. The filling with earth of the excavations after the concrete and masonry work below finish grade is completed.

Backfill, Hand. The act or process of placing excavated earth in a trench or back against the foundation of a structure, by a man and a hand shovel.

Backfire. In welding, a short pop of the flame from the torch tip followed by immediate reappearance or complete extinguishment of the flame.

Backflow. The unintentional flow of water into the supply pipes of a plumbing system from a non-supply source; back-siphonage is one type of backflow, generally due to a temporary occurrence of negative pressure (suction) in the pipes.

Backflow Connection. Any arrangement of pipe or fixtures which can cause backflow to occur.

Backflow Preventer. Backcheck valve; a device which prevents sewage from flowing backward into a building's plumbing system.

Backfurrow. The first cut of a plow from which the slice is laid on undisturbed soil.

Back Gouging. The forming of a bevel or groove on the other side of a partially welded joint to assure complete penetration upon subsequent welding from that side.

Background Noise. The sound level present in a room or space at any given time above which speech, music, desired signal, or sound must be raised in order to be heard or made intelligible.

Backhand Welding. A gas-welding technique wherein the flame is directed opposite to the progress of welding.

Backhoe. An excavating machine with a bucket rigidly attached to a hinged stick on a boom that is drawn toward the machine in operation.

Backhoe/Loader. An excavation machine combining a bucket on a hinged stick on a boom on one end, and a bucket or scoop at the other.

Backing. 1. Something forming a back. 2. The wood blocking behind plastering for supporting the load of lighting fixtures, cabinets, and hardware attached to a wall or ceiling. 3. Any material used as a base over which a finished material is to be installed. 4. See Plaster Base. 5. Material (metal, weld metal, asbestos, carbon, granular flux, etc.) backing up the joint during welding to facilitate obtaining a sound weld at the root. 6. The carpet foundation of jute, kraftcord, cotton, rayon, or polypropylene yarn that secures the pile yarns and provides stiffness, strength and dimensional stability. 7. The part of the wall behind the face brick.

Backing Board. A gypsum board product designed for use as the first or base layer in a multilayer system; also called Backer Board.

Backing Filler Material. Filler metal in the form of a ring, strip or consumable insert, fused in a single-welded joint.

Backing Off. See Featheredging tile.

Backing Ring. Backing in the form of a ring, generally used in the welding of piping.

Backing Rod. A foam plastic rod inserted in a joint to be sealed, to regulate the depth of sealant.

Backing Up. Constructing the inside section of a brick wall after the facing brick of the same wall has been laid header high.

Backlog. Reserve of uncompleted work.

Back-Mopping. Mopping the back or underside of roofing.

Back Mounted Tile. Mounted tile with perforated paper, fiber mesh, or other suitable bonding material applied to the backs or edges of the tile so that a relatively large portion of the tile area is exposed to the setting bed.

Backnailing. In roofing, the practice of blind nailing (in addition to hot- mopping) all the plies to a substrate to prevent slippage on slopes of 1-1/2 inch or more for steep asphalt, 1/2 inch or more for coal-tar pitch and dead-level asphalt.

Back-Plastering. Plaster applied to one face of a lath system following application and subsequent hardening of plaster applied to the opposite face; used primarily in construction of solid plaster partitions and certain exterior wall systems.

Back Pressure. 1. In plumbing, a higher air pressure at the sewer side of the trap than on the fixture side of the trap; water can, under certain conditions, be forced out of the trap in the wrong direction towards the fixture. 2. In a refrigerating system, the pressure on the low side of the system, also called suction pressure or low-side pressure.

Back Priming. A coat of paint applied to the back of woodwork and exterior siding to prevent moisture from entering the wood and causing the grain to swell.

Back Saw. Small toothed, joint-cutting saw with a metal back strip to keep the saw rigidly in line.

Back Seam. While all carpet seams are located on the back or underside of the carpet, those made when the carpet is turned over or face-down are called back seams, while those made with the carpet face-up are called face seams.

Backset of a Hinge. The distance from the edge of the door to the hinge.

Backset of a Lock. The distance from the centerline of a tubular door lock or cylinder to the edge of the door, measured on the high side of a beveled door.

Back Siphonage. The flow of water or other liquids, mixtures, or substances into the distributing pipes of a potable supply of water from any sources other than its intended course, due to a negative pressure in such pipe.

Back Splash. A protective panel installed on the wall behind a counter, sink or lavatory.

Backstep Sequence. A longitudinal sequence wherein the weld bead increments are deposited in the direction opposite to the progress of welding the joint.

Backstop. To support or bolster.

Backup. That part of a masonry wall behind the exterior facing.

Backup Bar. A small rectangular strip of steel applied beneath a joint to provide a solid base for beginning a weld between two steel structural members.

Backup Block. Concrete block which is used as a non-exposed structural wall and backs a finished surface to provide a complete wall system.

Back Vent. A branch vent installed for the purpose of protecting fixture traps from siphoning; back vents include most of the vents not installed specially to permit circulation between vent stacks and soil or waste stacks.

Back Wall. The wall facing an observer who is standing at the entrance to a room, shower, or tub shower.

Backwash. The process of cleaning the swimming pool filter by reversing the water flow.

Backwash Pipe. See Filter Waste Discharge Piping.

Bacteria. Microorganisms present in all water supplies including swimming pools; chlorine and other chemicals are used to keep these microorganisms under control.

Bacteriacide. Any of a number of chemicals used to kill bacteria.

Bad Debt. An uncollectable debt.

Baffle. Plate or vane used to direct or control movement of fluid or air within a confined area.

Bain-Marie. A kitchen utensil that has a hot water compartment under an upper compartment that keeps food warm; also called steam table.

Bake Oven. A cooking device containing a chamber wherein food is baked by dry heat; a commercial or institutional oven.

Bakelite. Trademark; any of various thermosetting resins or plastics made from formaldehyde and phenol.

Baking Finishes. Baking at elevated temperatures improves certain types of coatings used on metal articles, such as automobiles and refrigerators; baking may be done in an oven, under infrared lamps or by induction heating according to the demands of shape, space and other requirements; the article that is coated must, of course, be able to withstand the temperature required for the proper baking of the finish.

Baking Japan. An enamel to which the artificial heat of an oven is usually applied in order to attain the maximum hardness or toughness of film.

Balance. An instrument that measures mass, consisting of a central pivot, a horizontal beam, and two scales.

Balance Box. A loaded box at the far side of a crane from the jib and the load, to counterbalance them.

Balanced Construction. A method of constructing manufactured wood products so that moisture content changes will be uniformly distributed and therefore will not cause warping. An example would be symmetrical construction of plywood in which the grain direction of each ply is perpendicular to that of adjacent plies.

Balanced Cuts . 1. Cuts of tile at the perimeter of an area that will not take full tiles; the cuts on opposite sides of such an area shall be the same size. 2. The same sized cuts on each side of a miter.

Balanced Door. A door swung on an arm and pivot arrangement that is spaced out from the jamb so that some of the wind impinging on the door will assist in opening the door.

Balance Sheet. A statement of financial condition of a specific date.

Balancing Valve. A pipe valve that is used to control the flow of a liquid rather than to shut it off.

Balcony. A platform that projects from the wall of a building.

Baldachin. An ornamental canopy over an altar; also called baldachino.

Ball. An ornament of rounded form, common as the termination of a cupola or lantern.

Ballast. 1. A device used with a fluorescent and high intensity lamp, to provide the necessary circuit condition for starting and operation. 2. Any material used as non-structural fill or dead weight. 3. Heavy material, such as water, sand, or iron which has no function in a machine except increase of weight. 4. Crushed rock or gravel which is spread on a roof surface to form its final surface.

Ballast Replacement. 1. The replacement of broken stone, slag, and sand in railroads and highways to keep railroad ties in place and to provide drainage. 2. The replacement of the device that regulates current in a fluorescent lamp fixture.

Ball Bearing. A bearing in which the two halves are separated by a ring of small metal balls which reduce friction.

Ball Bearing Hinge. A hinge having ball bearings positioned between the protruding, cylindrical parts of the hinge, to prevent friction.

Ball Check Valve. Valve assembly (ball) which permits flow of fluid in one direction only.

Ball Clay. A secondary clay, commonly characterized by the presence of organic matter, high plasticity, high dry strength, long verification range, and a light color when fired.

Ball Cock. An automatic valve whose opening and closing are controlled by a spherical float at the end of a lever, such as on a flush toilet.

Ball Milling. A method of grinding and mixing material, with or without liquid, in a rotating cylinder or conical mill partially filled with grinding media such as balls or pebbles.

Balloon Frame. A wooden building frame composed of closely spaced members (studs) which are continuous from the sill to the top plate of the roof line; also called Eastern Frame; compare Platform Frame.

Balloon Payment. A large lump sum final payment to pay off an installment note, where the previous payments were much smaller.

Ball Peen Hammer. A hammer which has a rounded head at its opposite end, used extensively in machine shops.

Ballrace. Round or straight retainer to keep ball bearings in their proper operating position.

Ballroom. A large room or hall designed for dancing, balls, and similar festivities.

Ball, Tree. See Root Ball.

Ball Valve. A valve in which a ball regulates the opening by its rise and fall due to fluid pressure, a spring, or its own weight.

Baluster. Each of a series of often ornamental short posts or pillars supporting a rail or coping; also called banisters.

Balustrade. A row of balusters topped by a railing on a stair, porch, or balcony.

Bamboo. Giant tropical woody grasses with hollow stems that are used for making furniture.

Banc. The bench on which court judges sit.

Band. 1. A low, flat molding. 2. A group of bars distributed in a slab or wall or footing.

Band-Aid Approach. See Quick Fix.

Banding. Metal or plastic strapping to secure bundles of building products, such as gypsum wallboard, together in a shipping unit.

Band Joist. A wooden joist perpendicular to the direction of the joists in a floor framing system, closing off the floor platform at the outside face of the building.

Bandsaw. A machine saw with a narrow endless blade that runs over pulleys.

Band Stage. A raised platform where musicians perform.

Banister. A light baluster supporting a stair handrail.

Banjo Taper. A mechanical device which dispenses tape and taping compound simultaneously.

Bank. A building accommodating a financial establishment that receives and pays out money and deals generally in money and finance.

Bank, Duct. See Duct Bank.

Bank Gravel. Gravel found in natural deposits, usually more or less intermixed with fine material, such as sand or clay, or combinations thereof; gravelly clay, gravelly sand, clayey gravel, and sand gravel indicate the varying proportions of the materials in the mixture.

Bank Measure. The volume of earth in its natural site.

Bank Run Gravel. Excavated material that is generally 1/4 inch minimum to 6 inches maximum.

Bankruptcy. A legal proceeding by which a debtor may avoid legal and financial obligations.

Baptistry. A part of a church or formerly a separate building for baptism; also spelled baptistery.

Bar. 1. A straight wood or metal piece used as a lever, support, barrier, or fastening. 2. A barrier of any shape. 3. A submerged or partly submerged sand bank along a shore or in a river often obstructing navigation. 4. The railing in a courtroom separating the judge, jury, lawyers, and witnesses from the spectators. 5. A particular system of courts. 6. The whole body of lawyers qualified to practice in a jurisdiction. 7. A building, room, or counter at which food or beverages are served. 8. A small shop or stall serving refreshments, such as a snack bar. 9. A specialized department in a large store, as a watch repair bar. 10. A steel member used to reinforce concrete. 11. Unit of pressure; one bar equals 0.9869 atmospheres (one million dynes per square centimeter).

Barbed Nail. A nail with a barbed shank to resist withdrawal.

Barbed Wire. Wire that is twisted with barbs or sharp points.

Barbican. The outer defense of a city or castle, often a part of a system of barrier walls and consisting of a double tower above a gate or drawbridge.

Bar Chart. A simple construction scheduling technique which graphically shows the starting and finishing times for the various tasks which make up a job.

Bare Electrode. A filler-metal electrode used in arc welding consisting of a metal wire with no coating other than that incidental to the drawing of the wire.

Bare Solid Wire. Uninsulated single wire used as an electric conductor.

Bare Stranded Wire. Uninsulated group of fine wires used as a single electric conductor.

Barge. A floating platform or vessel from which construction activities may be performed; often used in rivers to install bridge piers and also used extensively in waterfront construction.

Bargeboard. A board, sometimes ornamental, fixed to the end of a gable, running from ridge to eave, to hide the ends of the roof timbers.

Bar Hanger. A metal bar, either straight or offset, to allow for the mounting of a ceiling outlet box between ceiling joists, or an outlet box, or switch box between wall studs; an adjustable hanger is one that can be made to fit the varying distances between floor and ceiling joists or rafters to securely hold electrical outlet boxes and devices.

Bar Hook. A semi-circular (180 degree) or a 90 degree turn at the free end of a steel reinforcing bar to provide anchorage in concrete.

Barium Sulphate. Heavy, white, extender pigment made from the mineral, barite; unaffected by acids, alkalis.

Bar Joist. A truss-like floor joist or rafter fabricated from steel bars.

Bark. The tough outer layer of a tree.

Bark pocket. An opening between a tree's annual growth rings that contains bark.

Bar Lap. The amount steel reinforcing bars must lap in order to develop sufficient bond to transfer their full load capacity.

Barn. 1. A large farm building for storing hay and grain and housing farm animals and equipment. 2. An unusually large and bare building, pejoratively. 3. A large building for storing a fleet of trucks or trolleys.

Bar Number. A number designating the size of a steel reinforcing bar, approximately the bar diameter in eighths of inches; for example, a #5 bar is approximately 5/8 inch in diameter; a #9 bar is approximately 1-1/8 inch in diameter (9/8); bar numbers are rolled onto the bar for easy identification.

Barometer. An instrument for measuring atmospheric air pressure, it may be calibrated in pounds per square inch, or in inches or millimeters of mercury in a column.

Baroque. Highly ornate, grotesque, flamboyant, and extravagant in style, especially of the architecture, art, and music of the 17th and 18th centuries.

Bar, Parallel. A pair of bars on a support, adjustable in height and spacing, that are parallel to each other, and are used for gymnastic exercises.

Barracks. 1. A building or building complex used to house soldiers. 2. A large plain building.

Barratry. The persistent incitement of litigation.

Bar, Reinforcing. See Reinforcing Bar.

Barrel. A unit of weight for cement. 376 lbs. net, equivalent to 4 U. S. bags of portland cement.

Barrel Ceiling. A rounded or semi-circular ceiling.

Barrel Shell. A scalloped roof structure of reinforced concrete that spans in one direction as a barrel vault and in the other as a folded plate.

Barrel Tile. A type of ceramic or cast roofing tile.

Barrel Vault. A segment of a cylinder that spans as an arch; used as a structural technique to support a ceiling or roof by having all of the components act in compression as an arched ceiling; used extensively in ancient buildings and into the 19thcentury, because no structural steel or timber is needed.

Barricade. A barrier; an obstruction to prevent passage or to prevent access.

Barrier. A fence or other obstacle that bars advance or access.

Barrier Coating. 1. Shielding or blocking coating or film. 2. See Transition Coating.

Barrier, Vapor. A type of plastic sheeting that both eliminates drafts and keeps moisture from damaging a building or structure.

Barroom. A room or establishment in which the principal feature is a bar for the serving of liquor.

Bar Spacing. Distance between parallel reinforcing bars measured from center to center of the bars perpendicular to their longitudinal axis.

Bar Support. A rigid device used to support or hold reinforcing bars in proper position to prevent displacement before or during concrete placement.

Bar Supports. Devices of formed wire, plastic or concrete, to support, hold and space reinforcing bars.

Basalt. A dark basic volcanic rock whose strata sometimes form columns.

Basalt Ware. A black unglazed vitreous ceramic ware having the appearance of basalt rock.

Base. 1. A molding or board placed along the bottom of a wall next to the floor; also called Baseboard. 2. A vinyl, wood or metal trim applied at the floor line to protect the vertical wall from damage. 3. The bottom of a column. 4. One or more rows of tile installed above the floor. 5. See Plaster Base. 6. The lowest part, or the lowest main division, of a building, column, pier, or wall. 7. The lowest point of any vertical pipe.

Baseball Backstop. A construction, often of wood or chain-link fencing, that encloses the home plate area on three sides and from the top to prevent balls from leaving the field and hitting spectators.

Base Bead. See Base Screed

Base Bid. A stipulated construction sum based on the contract documents.

Baseboard. See Base.

Baseboard Heater. Heating strips that are installed at the juncture of the wall and floor and may be either recessed or surface-mounted; generally along the outside walls of rooms.

Baseboard Radiator. A heating unit installed along a baseboard in a building or structure, is usually hydronic (hot water or steam).

Base Cabinet. 1. Case, box, or piece of furniture which rests on the floor with sets of drawers or shelves with doors, primarily used for storage. 2. Floor-mounted cabinets, usually with a counter, sink or appliance installed.

Base Cap. A finish fitting for tile.

Basecoat. Any plaster coat or coats applied prior to application of the finish coat.

Base Coat Floating. The finishing act of spreading, compacting, and smoothing of the base coat plaster to a reasonably true plane.

Base, Column. The plate beneath a column that distributes the load.

Base Course. 1. The lowest row of masonry in a wall, pier, foundation, or footing. 2. In asphaltic concrete paving, the layer of material immediately beneath the surface or intermediate course; it may be composed of crushed stone, crushed slag, crushed or uncrushed gravel and sand, or combinations of these materials and may also may be bound with asphalt.

Base Flashing. 1. A waterproof membrane that is attached to a roof and bent up the side of a chimney. 2. A waterproof membrane that is attached at a joint between any vertical member and a roof.

Base Grouting. The injection of grout to fill voids in subfloor slabs, foundations or concrete slabs.

Base Layer. The first or interior layer of gypsum board applied in a multilayer system.

Base Line. The main traverse or surveyed line running through the site of proposed construction, from which property lines, street lines, and buildings are located and plotted on the plan.

Base Lines. Part of a governmental land surveying grid system covering the country, with base lines running east and west and meridians running north and south; townships are located with reference to a specific base line and meridian; for example, T3N R12W, MDBM is read Township 3 North, Range 12 West, Mount Diablo Base Line and Meridian.

Base, Manhole. The cast iron frame into which a manhole cover fits.

Basement. The part of a building that is wholly or partially below ground level; cellar.

Basement Soil. See Subgrade.

Base Metal. The metal to be welded, cut or brazed.

Base Molding. A strip of wood used to trim the upper edge of a baseboard.

Base, Mogul. A screw-in style base for an incandescent lamp of generally 300 watts or more.

Base of Structure. The level at which earthquake motions are assumed to be imparted to a building; this level does not necessarily coincide with the ground level.

Baseplate. A steel plate inserted between a column and the foundation, used to level the column and to spread the load of the column to a larger area of the foundation.

Base Ply. The first layer in a built-up roof.

Base Screed. A preformed metal screed with perforated or expanded flanges; provides a ground for plaster and separates areas of dissimilar materials.

Base Shear. Total shear force acting at the base of a structure.

Base Sheet. This is the first ply in some multi-ply built-up roofing membranes; it is usually a saturated felt or a coated felt.

Base Shoe Molding. See Base Shoe.

Base Shoe. Wood strip, with a curved or projecting surface, used to finish intersection of base and floor; also called Base Shoe Molding, Shoe Mold, or Carpet Strip.

Base, Stone. The beginning or starter stone of a fieldstone wall.

Base, Terrazzo. A subfloor slab or foundation using a flooring material made from marble or other stone chips set in portland cement and polished when dry.

Basic Lead Carbonate. A type of white lead pigment.

Basic Lead Sulphate. A type of white lead.

Basilica . 1. An ancient Roman public hall with an apse and colonnades, used as a law court and place of assembly. 2. A similar building used as a Christian church.

Basin. 1.A wide shallow open container. 2. A lavatory. 3. A hollow rounded depression. 4. Any sheltered area of water where boats can moor safely. 5. A round valley. 6. An area drained by rivers and tributaries.

Basin Wrench. A plumber's tool used for installing hard-to-get-at fittings.

Basis for Acceptance. The method of determining whether a lot of material is acceptable under given or accepted specifications.

Basket. Wire assembly to support and space dowel bars and expansion joints in concrete slabs on the ground.

Basketweave Bound. Module groups of brick laid at right angles to those adjacent.

Bas-Relief. Sculpture or carving in which the figures project only slightly from the background surface with no undercutting.

Bastard File. A course file for rough shaping of metal or wood.

Bastard Granite. A quarrier's term for nearly any stone which may not be considered a true granite, particularly applied to gneiss.

Bastard Sawn. Lumber, primarily hardwoods, in which the annual rings make angles of 30 to 60 degrees with the surface of the piece.

Bat. A broken brick; an end portion of a brick; approximately a half.

Batavia. Batavia dammar; see Dammar.

Batch Disposal. The orderly placement or distribution of freshly mixed concrete or masonry.

Batch Mixer. A machine which mixes batches of concrete or mortar in contrast to a continuous mixer.

Batch Plant. 1. A manufacturing facility for producing asphalt paving mixtures that proportions the aggregate constituents into the mix by weighed batches and adds asphalt material by either weight or volume. 2.An operating installation of equipment including batchers and mixers as required for batching or for batching and mixing concrete materials; also called mixing plant when equipment is included.

Bath. 1. A tub used for bathing. 2. The room that contains the tub. 3. Liquid solution used for cleaning, plating, or maintaining a specified temperature.

Bath Accessory. Bathroom equipment such as towel bars, grab bars, grab rails, soap dishes, toilet paper holders, and medicine cabinets.

Bath Chair. A wheelchair.

Bath Faucet. A valve used to draw hot or cold water into a bathtub.

Bathhouse. A building with baths and dressing rooms for public use.

Bathroom . 1. A room containing a bath tub or shower and usually a toilet and lavatory. 2. A polite term for any toilet room.

Bathroom Lock. A non-keyed privacy lock that can be locked from the inside by push-button and opened from the outside by a small tool; a privacy lock.

Bath Tub Door. A folding or sliding door mounted on a bath tub rim to keep water spray within the tub area when there is a shower over the tub.

Bath Tub. The tub in a bath room.

Bath Vanity. A bathroom cabinet with a lavatory mounted in the counter top.

Batt Insulation. Fiber or wool insulation in sheet form, usually with a paper lining; a preformed section of inorganic insulation sized to fit snugly in a framed cavity.

Batten. 1. A thin narrow strip of wood or metal to cover or reinforce a joint; a predecorated strip or joint covering designed to conceal the junction between adjacent boards; frequently used in demountable systems. 2. A cleat; a narrow strip of board used to fasten several pieces together.

Batten Seam. A seam in a sheet metal roof.

Batten Siding. Vertical siding which has narrow strips of metal or wood covering the joints.

Batter. 1. A slight receding upward slope to the outward face of a building wall or a retaining wall. 2. Stepping or sloping masonry back in successive courses; the opposite of corbel.

Batter Board. A temporary framework used to assist in locating the corners when laying a foundation; also used to maintain proper elevations of structures, excavations and trenches in any kind of below ground construction.

Batter Pile. Pile driven at an angle to brace a structure against lateral or horizontal thrust.

Battery. One or more cells connected together to furnish electric current by interaction of metals and chemicals.

Battery of Fixtures. Any group of two or more similar adjacent plumbing fixtures which discharge into a common horizontal waste or soil branch.

Battlement. A parapet with recesses along the top of a wall, originally as a fortification, now as ornamentation.

Baudelot Cooler. Heat exchanger in which water flows by gravity over the outside of the tubes or plates.

Bauhaus. An architectural school in Germany, founded by Walter Gropius in 1923, emphasizing technology, craftsmanship, and design aesthetics.

Bauxite. A claylike mineral containing varying proportions of alumina, the principal source of aluminum.

Bay. One of the intervals or spaces into which a building plan is divided by columns, piers, or division walls.

Bay Window. A rectangular, curved, or polygonal window projecting out from the face of a wall.

Bazaar. 1. A market in an oriental country consisting of rows of shops or stalls offering miscellaneous goods. 2. A large shop selling fancy goods. 3. A department store.

Bead. 1. A narrow line of weld metal or sealant. 2. A strip of metal or wood used to hold a sheet of glass in place. 3. A narrow, convex molding profile. 4. A metal edge or corner accessory for plaster or gypsum board.

Beaded Moulding. A cast plaster string of beads planted in a moulding or cornice.

Beading. A ridge or raised linear deformation that may appear at finished gypsum board joints.

Bead Weld. A type of weld composed of one or more string or weave beads deposited on an unbroken surface; the metal added in welding.

Bead, Molding. 1. A half-round narrow molding, attached or milled on a larger piece. 2. A square or rectangular trim less than 1 inch in width.

Bead, Parting. A narrow strip between the upper and lower sashes in a double-hung window frame.

Bead, Plaster. Built-in edging usually metal, to strengthen a plaster angle.

Bead, Sash. A strip with one edge molded, against which a sash slides.

Beam. A straight structural member that acts primarily to resist transverse loads; a structural element which sustains transverse loading and develops internal forces of bending and shear in resisting the loads; an inclusive term for joists, girders, rafters, and purlins.

Beam Bolster. A fabricated wire device used to temporarily support reinforcing steel in structural formwork such as beams and slabs.

Beam, Bond. A reinforced horizontal concrete masonry or concrete beam installed in place to strengthen a masonry wall and tie a masonry wall together.

Beam Chair. A wire seat or support for reinforcing bars designed to maintain their location while concrete for a beam is poured around them.

Beam Clamp. A device which holds a horizontal structural member to a vertical member.

Beam Compass. A drafting tool consisting of a wood or metal bar fitted with a point and a movable pen or pencil holder to use in drawing large radius arcs and circles; a Trammel.

Beam, Concrete. A horizontal structural member which transversely supports a load and transfers the load to vertical members, made of a composite material consisting of sand, coarse aggregate, cement and water.

Beam Fireproofing. Fire-resistant materials that cover a horizontal structural member, to insure structural integrity in the event of a fire.

Beam Formwork. The system of support for freshly placed concrete for a horizontal structural member.

Beam Furring. Strips of wood or metal fastened to a horizontal structural member to form an airspace, to give the appearance of greater thickness, or for the application of an interior finish such as plaster.

Beam, Grade. An end-supported horizontal load-bearing foundation member that supports an exterior wall or other building load.

Beam Hanger. 1. A strap wire or other device used to hang beam forms from another structural member. 2. A steel strap used for securing the end of a beam.

Beam Plaster. A combination of cementitious material and aggregate mixed with water and applied to a horizontal structural member, preserving in a rigid state the form or texture imposed during the period of elasticity.

Beam, Precast. A concrete horizontal structural member that is cast and cured in other than its final position, on- or off-site.

Beam Rail. A solid wood band attached to a horizontal structural member.

Beam, Reinforcing. A horizontal member installed to strengthen and support the load of a structure.

Beam, Sheetrock. A horizontal member constructed of plasterboard.

Bearding. In carpeting, long fiber fuzz occurring on some loop pile fabrics, caused by fibers snagging and loosening due to inadequate anchorage.

Bearing. 1. Part of a machine that supports and aligns a rotating or other moving part. 2. That part of a lintel, beam, girder or truss, which rests upon a column, pier or wall.

Bearing Block. A piece of wood fastened to a column to provide support for a beam or girder.

Bearing Capacity. Allowable bearing capacity; the maximum allowable load on a structural element.

Bearing Pad. A block of metal, plastic, or synthetic material used to cushion the point at which one structural element rests upon another.

Bearing Partition. A partition that supports any vertical load in addition to its own weight.

Bearing Plate. A plate placed under a truss, beam, girder, or column to distribute the load.

Bearing Wall. A wall which supports any vertical loads in addition to its own weight.

Beating Block. A wooden block used to embed tiles in a flat plane; the method used is called beating in.

Becquerel. A unit for measuring radiation from radium and radon.

Bed. 1. The horizontal surface on which the bricks of a wall lie in courses, also the mortar on which the brick rests. 2. See Bed Mold.

Bed Coat. See Bedding Coat.

Bedding. 1. A filling of mortar, putty, or other substance in order to secure a firm bearing. 2. Ground or supports in which pipe is laid.

Bedding Coat. 1. In gypsum board, the first coat of joint compound over tape, bead, and fastener heads; also called Bed Coat. 2. That coat of plaster to receive aggregate or other decorative material of any size, impinged or embedded into its surface before it sets.

Bedding Plane. A separation or weakness between two layers of rock, caused by changes during the building up of the rock-forming material.

Bed Joint. The horizontal layer of mortar in which a masonry unit is set.

Bed Mold. A flat area in a cornice, designed to have enrichment planted later; also called Bed.

Bed Molding. A molding in an angle, as between the overhanging cornice or eaves of a building and the sidewalls.

Bedrock. A solid layer or stratum of rock beneath ground level; solid rock as distinguished from boulders.

Bedroom. A room for sleeping.

Beeswax. Wax produced by honey bee.

Behavioral Aspects of Management. These include human relations, cooperation, responsibility, communication and executive ability.

Belfry. A bell tower.

Bell. That portion of a pipe which, for a short distance, is sufficiently enlarged to receive the end of another pipe of the same diameter for the purpose of making a joint; also called Bell End or Hub.

Bell and Spigot Joint. A type of joint used in cast iron pipe where a male pipe end (the spigot) slips into a female end (the bell) and is then caulked with oakum and sealed with lead.

Bell End. See Bell.

Belleek China. A highly translucent whiteware composed of a body containing a significant amount of grit and normally having a luster glaze.

Bellows Copper. A flexible joint in copper piping that can expand or contract to allow for thermal fluctuations.

Bellows Seal. Method of sealing the valve stem; the ends of the sealing material are fastened to the bonnet and to the stem; the seal expands and contracts with the stem level.

Bellows. An instrument or machine that draws in air through a valve or orifice by expansion and contraction and expels it through a tube.

Bell Trap. A floor drain trap consisting of an inverted bell with a water seal.

Belly-Up. The condition of a contracting firm when it becomes insolvent and goes bankrupt.

Belt Course. 1. A course of bricks or other material projecting slightly from the face of the wall. 2. A horizontal board across or around a building, usually made of a flat member and a molding. 3. A horizontal course on the face of a building; when continuous with a row of window sills or lintels, is referred to as a Sill Course or a Lintel course; also called a String Course.

Belt Marks. A surface defect in gypsum board made by the machine's forming belt during manufacture.

Belt sander. An electric sanding tool where the sand paper abrasive is a continuous belt.

Bema. The part of an eastern church containing the altar.

Bench. 1. A long seat for seating several people. 2. A judge's seat in a lawcourt. 3. A working table or counter for a carpenter, artisan, or scientist. 4. A low scaffold board that allows the gypsum board hanging crew to easily reach the ceiling area; also called Hanger's Bench.

Bench Mark. A point of known or assumed elevation used as a reference in determining and recording other elevations.

Bench Saw. A power saw held securely on a stationary bench.

Bend, Soil. A piece of short, curved pipe, like an elbow, used to connect two straight links of pipe in a sewage system.

Bending. The result of a force which tends to cause curvature in a linear element; internal stresses of tension and compression are a result of this action.

Bending Moment. The sum of moments for all forces that occur above the neutral axis; the moment that causes a beam or other structural member to bend; see Moment.

Bending Spring. Coil spring which is placed on inside or outside of tubing to keep it from collapsing while bending it.

Bending Stress. A compressive or tensile stress resulting from the application of a nonaxial force to a structural member.

Bends, The. Caisson disease

Beneficial Occupancy. The use of a premises or a portion for the uses intended, although the work of the project may not be completed.

Beneficiary. 1. The person designated to receive the proceeds of an insurance policy. 2. In a trust deed, the lender.

Benny. Slang for benzine.

Bent. A plane of framing consisting of beams and columns joined together, often with rigid joints; a single vertical framework consisting of horizontal and vertical members supporting the deck of a bridge or pier.

Bent Bar. A steel reinforcing bar bent to a prescribed shape such as a truss bar, straight bar with hook, stirrup, or column tie.

Bentonite. A clay composed principally of minerals of the montmorillonoid group, characterized by high absorption and very large volume change with wetting or drying, commonly swelling to several times its dry volume when saturated with liquid.

Bentonite Membrane. A thin, pliable sheet or layer manufactured from bentonite; used for waterproofing retaining walls.

Benzene. A material derived from coal tar and widely used a solvent; also used for cleaning after painting and other finishing operations; a very powerful aromatic solvent for many materials, its use is restricted because of its toxicity and because it is a fire hazard; often confused with Benzine due to similarity in pronunciation; also called Benzol.

Benzine. A highly flammable petroleum product used by painters as thinning solvent and diluent; also used in dyeing and as a cleansing agent; also called Varnish Makers' and Painters' Naphtha (VM&P).

Benzol. See Benzene.

Berm. A small artificial ridge of soil at the top of an earth bank to prevent water drainage onto the bank.

Bernoulli's Theorem. In a stream of liquid the sum of elevation head, pressure head, and velocity remains constant along any line of flow provided no work is done by or upon the liquid in the course of its flow and decreases in proportion to energy lost in flow.

Beryllium Oxide. An inorganic material of exceptionally high thermal conductivity which is toxic in the powder form; BeO; also called Berylla).

Bespoke. Custom-made.

Bevel. 1. A slanted surface; an end or edge cut at an angle other than a right angle. 2. A type of edge preparation on metal where welding is to take place.

Bevel Board. A board used in framing roof or stairway to lay out bevels; also called Pitch Board.

Beveled Concrete. An angle in concrete or inclination of any line in concrete or concrete surface that joins another.

Beveled Edge. The factory edge of gypsum board that has been angled to form a vee grooved joint when two pieces are placed together.

Bevel Gears. Meshing gears on intersecting axes.

Beveling. A type of chamfering.

Bevel of a Door. The angle of the lock edge in relation to the face of the lock stile; standard bevel is 1/8 inch in 2 inches (3.2 mm in 50.8 mm).

Bevel Siding. A type of finish siding used on the exterior of a house; usually manufactured by resawing a dry, squared, surfaced board diagonally to produce two wedge- shaped pieces; also called Lap Siding.

Bezel. 1. A sloping surface on a cutting tool, as on a chisel. 2. A frame around something like a glazed opening or a built-in appliance.

BHMA Hinge. A hinge that has the seal of approval of the Builders Hardware Manufacturers' Association.

Bib. See Hose Bibb.

Bibb, Hose. See Hose Bibb.

Bicycle, Exercise. A stationary bicycle used for health and fitness purposes in gyms or health clubs.

Bid. An offer to perform; an offer to enter into a contract usually for a stipulated sum of money; a tender; a proposal.

Bid Bond. A bond, secured by a bidder, which guarantees that the bidder selected by the owner will accept the project, or the owner will have the project for the bid price as noted in the accepted bid.

Bid Date. See Bid Time.

Bidder. A prime contractor who submits a bid directly to an owner.

Bidding Documents. The bidding requirements and the proposed contract documents, including any addenda issued prior to receipt of bids.

Bidding or Negotiation Phase. One of the standard phases of architectural service (Schematic Design Phase, Design Development Phase, Construction Documents Phase, Bidding or Negotiation Phase, and Construction Phase-Administration ot the Construction Contract).

Bidding Period. The time period during which contractors can prepare their bid proposals.

Bidding Requirements. The group of documents issued to general contractors who are bidding a construction project, other than the contract documents; bidding requirements include the Advertisement or Invitation to Bid, Instructions to Bidders, Supplementary Instructions to Bidders, the bid form, and other sample bidding and contract forms.

Bidet. A plumbing fixture about the size and shape of a toilet, equipped with cold and hot running water, used for personal hygiene.

Bid Package. The group of documents issued to general contractors who are bidding a construction project, consisting of the bidding requirements and the contract documents; also called the Bidding Documents.

Bid Rigging. Collusion; a fraudulent practice involving secret agreements between or among competing contractors or owners' agents to control the outcome of a bidding procedure.

Bid Shopping. A procedure wherein a general contractor contacts subcontractors in an attempt to receive lower subcontractor prices after having beer awarded the contract for the project; considered unethical by the building industry.

Bid Time. The date and time designated by the architect and owner for the receiving of bid proposals; also called Bid Date.

Bi-Fold Door. A door with two leaves, hinged together to close on itself; one edge of each leaf is hinged at the jamb and the other edge is connected and guided by an overhead track.

Bifunctional. Having two functions.

Bike Rack. A framework installed on a site to store bicycles for the building's inhabitants, commonly specified in educational complexes.

Billhook. A tool with a hooked blade used for cutting and pruning.

Bimetallic Strip. A strip made of two metals that bends as the temperature changes; a bimetallic strip of iron and brass, when heated, will bend, as brass expands more than iron, thus bending the strip, and unbending upon cooling; this action can be used to open or close a switch.

Bin. A box, container, or enclosed place for storage.

Binder. 1. A substance that acts cohesively, see Cementitious Material. 2. Fines which hold gravel together when it is dry. 3. The nonvolatile portion of a paint which serves to bind or cement the pigment particles together; oils, varnishes and proteins are examples of binders; see vehicle. 4. A chemical additive to gypsum board core formulation, frequently starch, to improve the bond between the surfacing papers and the core.

Bindery. A building, workshop or factory where books are bound.

Binding. A strip sewed over a carpet edge for protection against unraveling.

Bin Method. A method of computing cooling energy use requirements for commercial and industrial building with unusual operating needs and for residences utilizing passive heating/cooling design with high mass thermal storage.

Bin, Ice Storage. A box, frame or enclosed place used for the storage of ice and the maintaining of its solid properties.

Biodegradability. The ability of a material to rot away after use, by the action of microorganisms.

Biological Contaminants. Agents derived from or that are living organisms, for example, viruses, bacteria, fungi, and mammal and bird antigens, that can be inhaled and can cause many types of health effects including allergic reactions, respiratory disorders, hypersensitivity diseases, and infectious diseases; also referred to as microbiologicals or microbials.

BIPV. Building Integrated Photovoltaic Cell.

Birch. Hardwood, species Betula.

Birch Door. A door constructed or faced with a strong fine-grained birch hardwood.

Birch Paneling. Rectangular sheets of paneling that have been constructed of a strong fine-grained birch hardwood.

Birch Veneer. Thin sheets of strong fine-grained hardwood used in furniture, flooring, and building paneling.

Birdbath. A hollow rounded depression in asphaltic concrete paving or in portland cement concrete flatwork; see Ponding, 1.

Birdcage. Colloquial name for the end of a stair rail where the banisters are curved in a spiral to form a newel post.

Bird Peck. A small hole or patch of distorted grain in wood resulting from birds pecking through the growing cells in the tree.

Birdscreen. Wire screening attached to louvers, ventilators and openings in a building or structure to prevent birds and small animals from entering.

Bird's-Eye. Small, localized area in wood where the fibers are indented and otherwise contorted to form small circular or elliptical figures which look somewhat like birds' eyes; common in sugar maple and used for decorative purposes; rare in other hardwood species.

Bird's-Mouth. A notch cut on the underside of a rafter to fit it to the top plate; not a full notch if there is no rafter overhang.

Biscuit Chips. Glazed-over chips on the edge or corner of the body of a tile.

Biscuit Cracks. Any fractures in the body of a tile visible both on face and back.

Bisect. 1. To divide into two parts. 2. To divide into two equal parts.

Bisque Fire. The process of kiln-firing ceramic ware prior to glazing.

Bit. 1. A unit of computer information expressed as a choice between two possibilities. a 0 or 1 in binary notation; short for binary digit. 2. Of a key, the projecting blade, cut in a manner that actuates the lock tumblers. 3. The removable cutting edge of a tool, as a drill bit. 4. Of a screw gun, the replaceable tip or portion that seats in the slotted screw head.

Bite. The amount of the edge of a pane of glass that is covered by the stop.

Biting. Solvent in topcoat of a lacquer surface dissolves or bites into coat below; if lacquer solvent is too biting, the dried lacquer surface may be rough or produce an orange peel effect.

Bitting. The configuration of notches and hollows on the blade of a door lock key.

Bitumen . A class of black or dark-colored solid, semisolid, or viscous cementitious substances, natural or manufactured, composed principally of high molecular weight hydrocarbons, of which asphalts, tars, pitches and asphaltites are typical; a generic term for an amorphous, semi-solid mixture of hydrocarbons derived from petroleum or coal; in the roofing industry there are two basic bitumens. coal-tar pitch and asphalt; before use, the bitumens are heated to a liquid state, dissolved with a solvent, or emulsified.

Bituminous. Resembling, containing, or impregnated with various mixtures of hydrocarbons, like tar or asphalt, together with their nonmetallic derivatives.

Bituminous Dampproofing. Impregnated mixtures of hydrocarbons, like tar or asphalt, together with their nonmetallic derivatives used on a surface to prevent moisture from passing through; dampproofing will not ordinarily resist water under hydrostatic pressure; see Dampproofing.

Bituminous Emulsion. 1. A suspension of minute globules of bituminous material in water or in an aqueous solution. 2. Invert emulsion; a suspension of minute globules of water or an aqueous solution in a liquid bituminous material.

Bituminous Grout. A mixture of bituminous material and fine sand that will flow into place without mechanical manipulation when heated.

Bituminous Membrane. A thin layer or sheet of an impregnated mixture of hydrocarbons together with their nonmetallic derivatives; used as dampproofing.

Bituminous Sidewalk. A walkway constructed with an impregnated mixture of hydrocarbons together with aggregate such as sand or stone; commonly called blacktop.

Black Iron Cap. A steel fitting, with female threads, which seals the end of a pipe.

Blackboard. A panel or board with a smooth, usually dark, surface for writing on with chalk.

Blacktop. A term used in the trade for asphaltic concrete paving.

Blade. 1. The flattened part of a paddle or oar. 2. The projecting parts of a fan, turbine, or propeller. 3. The sharp cutting part of a knife or saw. 4. The flat or concave part of a bulldozer. 5. See Steel Square.

Blade Grader. A grader.

Blaine Fineness. The fineness of powdered materials such as cement and pozzolans, expressed as surface area usually in square centimeters per gram, determined by the Blaine air-permeability apparatus and procedure.

Blanc Fixe. Artificially prepared barium sulphate; an extender pigment.

Blanket. Soil or broken rock left or placed over a blast to confine or direct throw of fragments.

Blanket Insulation. Thermal insulating material made of fibrous glass or mineral wool, sometimes with paper or foil surfacing, formed in batts or rolls.

Blanket Mortgage. A single trust deed or mortgage that covers two or more parcels of real property.

Blast Angle. Angle of sand blasting nozzle with reference to surface.

Blast Cleaning. Cleaning with propelled abrasives.

Blast Freezer. Low-temperature evaporator which uses a fan to force air rapidly over the evaporator surface.

Blast Furnace Slag. A non-metallic waste product developed in the manufacture of pig iron, consisting basically of a mixture of lime, silica and alumina, the same oxides that make up portland cement, but not in the same proportions or forms; it is used both in the manufacture of portland blast furnace slag cement and as an aggregate for lightweight concrete.

Blast Furnace. A smelting furnace into which compressed hot air is driven.

Bleach. To remove color or stains from.

Bleacher. An outdoor tiered stand of benches to provide seating at a sports ground.

Bleacher, Gym. An uncovered series of tiered planks used to seat spectators in a school or public gymnasium.

Bleaching. Restoring discolored or stained wood to its normal color or making it lighter by using bleaching agents.

Bleb. A small blister or bubble.

Bleeder Gun. A spray gun with no air valve; trigger controls fluid flow only.

Bleeding. 1. The upward movement of asphalt in an asphalt pavement resulting in the formation of a film of asphalt on the surface; usually caused by too much asphalt in one or more of the pavement courses, resulting from too rich a plant mix, an improperly constructed seal coat, too heavy a prime or tack coat, or solvent carrying asphalt to the surface; usually occurs in hot weather; also called Flushing. 2. The autogenous flow of mixing water within, or its emergence from newly placed concrete or mortar; caused by the settlement of the solid materials within the mass; also called Water Gain. 3. In gypsum board, a discoloration, usually at a joint, which may occur on a finished wall or ceiling. 4. Penetration of color from the underlying surface; seeping of a stain or lower coat through the top coat, spoiling the appearance of the top coat. 5. When coloring material from the wood or undercoat works into succeeding coats and imparts to them a certain amount of color; see Extractive Bleeding. 6. Slowly reducing the pressure of liquid or gas from a system or cylinder by slightly opening a valve. 7. The draining or loosening of saturants from the roofing material.

Bleeding Stain. Stain which works up or bleeds through succeeding coats of finishing materials.

Bleed-Valve. Valve with a small opening inside which permits a minimum fluid flow when valve is closed.

Blemish. A physical imperfection that affects appearance.

Blend. 1. To mix or make homogeneous. 2. A mixture, as of two pigments, to obtain a desired color.

Blending. Mixing one color with another so the colors mix or merge gradually.

Blind. 1. A panel, shade, or screen used on a window to block out light, give protection, add insulation, or as decoration. 2. A shutter.

Blind Header. A concealed header in the interior of a wall, not showing on the faces.

Blinding. Compacting soil immediately over a tile drain to reduce its tendency to move into the tile.

Blind Nailing. 1. Attaching boards to framing or sheathing with nails driven through the edge of each piece so as to be concealed by the succeeding board. 2. In installing tongue-and-groove flooring, the nails are placed at the root of the tongue where they will be hidden by the groove of the next piece; the nails pierce the subfloor at a 45 degree angle. 3. The practice of nailing the back portion of a roofing ply so that the nails will be concealed by the next ply of roofing.

Blind Stop. A rectangular wood molding, usually 3/4 by 1-38 inches, in a window assembly that receives the window screen frame.

Blind Story. A building story that has no windows.

Blind Vent. An ineffective, sometimes illegal, vent which stops in a wall thus giving the appearance of a vent but not actually functioning as a vent.

Blister. 1. A raised spot in a built-up roof caused by expansion of entrapped moisture, water vapor, or other gases, between any of the layers of roofing or mopping. 2. A loose raised spot on the gypsum board face usually due to an air space or void in the core.

Blistering. 1. In ceramic tile, the development, during firing, of enclosed or broken macroscopic vesicles or bubbles in a body or in a glaze or other coating. 2. A bulging of the finish plaster coat as it separates and draws away from the basecoat; the resulting protuberances are often termed turtle backs. 3. Formation of bubbles on surface of paint or varnish film, generally caused by moisture behind the film or excessive heat.

Blisters. 1. Cloudy or milky-looking raised spots on finished surfaces. 2. Protuberances on the finish coat of plaster caused by application over a too damp base coat or troweling too soon.

Block. 1. A compact, solid piece of substantial material that is worked or altered from its natural state to serve a particular purpose. 2. A wooden or metal case enclosing pulleys and having a hook, eye, or strap by which it may be attached. 3. A building divided into separate functional parts. 4. A part of a building or structure distinctive in some respect.

Block, Angle. A square of tile specially made for changing direction of the trim.

Block Coating. See Transition Primer.

Block, Concrete. A hollow concrete masonry unit constructed of a composite material consisting of sand, coarse aggregate, cement, and water.

Block Dampproofing. The act or process of applying a water-resistant material to the surface of a concrete or masonry block to prevent passage or absorption of water or moisture; see Dampproofing.

Block Demolition. The act or process of tearing down an old block wall structure.

Blocked Diaphragm. A diaphragm in which all sheathing edges not occurring on framing members are supported on and connected to wood blocking.

Block, Glass. A hollow masonry unit made of glass.

Block, Granite. A masonry unit consisting of a very hard natural igneous rock used for its firmness and endurance.

Block Grout. Mortar mixes used in block walls to fill voids and joints.

Blocking. 1. Pieces of wood inserted tightly between joists, studs, or rafters in a building frame to stabilize the structure, inhibit the passage of fire, provide a nailing surface for finish materials, or retain insulation. 2. A system of tying together two brick walls that were not built at the same time; the two adjoining or intersecting walls are tied together by offset and overhanging blocks of courses of bricks.

Blocking, Wood. Wood blocks used as filler pieces or stabilization between framing members.

Block Plane. Woodworking hand tool, for final smoothing.

Blocks. The usually rectangular areas of land between the streets in a town or city.

Block Sequence. A combined longitudinal and buildup sequence for a continuous multiple-pass weld wherein separated lengths are completely or partially built up in cross-section before intervening lengths are deposited; see Backstep Sequence.

Block, Splash. A small masonry block placed in the ground beneath a downspout to receive roof drainage and prevent standing water or soil erosion.

Block, Terminal. A decorative element forming the end of a block structure.

Block Vent. An opening serving as an outlet or inlet for air in a block structure.

Blood Analyzer. A mechanical apparatus which examines blood for medical or scientific purposes.

Blood Refrigerator. A mechanical cooling device or room used to keep stored blood at the proper temperature.

Bloom. 1. A visible exudation or efflorescence on the surface. 2. Whitening; blushing; clouded appearance on varnished surface. 3. Condition of clouding or fogging of paint film, usually caused by reactive materials in paint film coming into contact with dust, oil, deposits from gases in the air or soluble matter in rain.

Blots. Marks or stains on the face of a tile.

Blow. In gypsum board, a large area of paper separated from the core during the manufacturing process; it may appear as a large puffy blister or a full loose sheet of paper.

Blow Back. Rebound of atomized sprayed material.

Blown-In Insulation. Loose cellulose insulation that is blown into an attic, crawl space, or walls by a blowing machine.

Blown Oil. A vegetable or fish oil which has been thickened by air blown through it.

Blowout. A concrete form giving way.

Blowtorch. A gasoline torch used in burning off paint film; should be used only by experienced painters; it is a dangerous fire hazard when used by amateurs.

Blue Lead. A basic sulphate of lead containing small amounts of lead sulphide and carbon that impart a bluish-gray color; used primarily for its rust-preventive value.

Blueprint. 1.An obsolete method of copying construction drawings and maps; a wet process that produces a print with white lines on a blue background; a cyanotype or sun print. 2. Loosely, any construction drawing.

Blue Stain. A bluish or greyish discoloration of the sapwood caused by the growth of certain dark-colored fungi on the surface and in the interior of wood.

Bluestone. A sandstone of a dark-greenish to bluish-gray color that splits into thin slabs, commonly used to pave surfaces for pedestrian traffic.

Blue-Tops. Surveyor's stakes, marked with a blue lumber marking crayon, that should not be disturbed.

Blunder. A mistake or error made through clumsiness, stupidity, ignorance, or carelessness.

Blunging. The wet process of blending or suspending ceramic material in liquid by agitation.

Blushing. 1. A finish is said to blush when it takes on white or grayish cast during drying period; usually caused by the precipitation or separating of a portion of the solid content of the material, causing an opaque appearance. 2. Applied to lacquers when they become flat or opaque and white on drying; usually occurs when applied in a humid atmosphere.

BM. Bench mark.

Board. 1. A flat thin piece of sawn lumber, usually long and narrow; lumber less than 2 inches thick. 2. A group of persons having supervisory, investigatory, or advisory powers. 3. A sheet of insulating material carrying circuit elements and terminals so that it can be inserted in an electronic apparatus. 4. A flat panel of compressed fibers. 5. A flat panel used for posting bulletins or as a chalkboard.

Board and Batten. Linear vertical wood boards with wood strips covering vertical joints used as an exterior cladding for a framed wall.

Board Cement. An adhesive used to bond pieces of wood together.

Board, Composite. A board that is made of several compressed materials; used for sheathing, wallboard, or as an insulation or acoustical barrier.

Board, Concrete Finish. Wooden boards placed in the concrete formwork as for liners to provide a wood-pattern finish to the completed reinforced concrete.

Board, Diving. A narrow platform, often equipped with a spring and raised off the ground by a ladder, attached to the edge of a swimming pool and extending over the water, used to give divers altitude and lift.

Board, Dock. Heavy timber used in the construction of the raised platform used for the loading and unloading of trucks.

Board Foot. A unit of lumber volume, a rectangular solid nominally 12" x 12" x 1".

Boarding In. The process of nailing boards on the outside studding of a house.

Board Insulation. Rigid form of plastic foam used as a rigid application insulating surface, commonly of polyurethane.

Board Knife. A hand tool holding a replaceable blade to sharply score or trim gypsumboard products; a popular brand is a Stanley knife.

Board Measure. The standard system for the measurement of lumber; see Board Foot.

Board Rafter. The wooden structural member that makes up the sloping parallel beams used to support a roof.

Board, Ridge. The board against which the tips of rafters are fastened; the top line of a roof; the ridge.

Board Saw. A short hand saw with very coarse teeth for cutting gypsumboard for door and window frame openings.

Board Siding. A type of lumber installed on the exterior walls of a building or structure to act as the finish sheathing.

Board Sub-Flooring. A wooden member that is installed on floor joists to which the finished floor is fastened.

Boatswain's Chair. A trapeze-like seat that is slung from rigging to support a worker; a Bosun's Chair.

BOCA. Building Officials and Code Administrators International, an organization that publishes model building codes.

Bodega. 1. A wine vault or cellar; a wine shop where wine is drawn from barrels. 2. A storeroom or warehouse.

Bodied Linseed Oil. Linseed oil that has been thickened by suitable processing with heat or chemicals; bodied oils vary greatly in viscosity; some are little thicker than raw linseed oil; others are almost jelly like.

Body. 1. The structural portion of ceramic articles; also refers to the material or mixture from which the article is made. 2. The structural portion of an article covered with ceramic tile. 3. Thickness, consistency, or viscosity of a fluid. 4. See Steel Square.

Body Coat. Intermediate coat of paint between priming and finishing.

Body Feed. The continuous addition of small amounts of filter aid during the operation of a diatomaceous earth filter.

Boiled Linseed Oil. Linseed oil to which enough lead, manganese, or cobalt salts have been added to make the oil harden more rapidly when spread in thin coatings; drying properties accentuated by heating oil to 130 to 200° C.

Boiled Oil. Drying oil treated with driers to shorten the drying time.

Boiler. 1. A fuel-burning apparatus for producing hot water or steam. 2. A heating system in which water is used as the distribution medium.

Boiler Bow Off. An outlet on a boiler to permit emptying or discharging of water or sediment in the boiler.

Boiler, High-Pressure. Boiler furnishing steam at pressures of 15 pounds per square inch gauge or higher (1.05 kg/cm^2).

Boiler Horsepower. Term now seldom used, meaning equivalent to a heating capacity of 33,475 Btu/hr. (983 watts).

Boiler, Hot-Water and Low-Pressure Steam. A boiler furnishing hot water at pressures not more than 30 pounds per square inch gauge (2.12 kg/cm^2) or steam at pressures not more than 15 pounds per square inch gauge (1.06 kg/cm^2).

Boiler Plate. Standardized, formulaic, or hackneyed language in a contract.

Boiler Room. The space provided for a hot water or steam boiler, circulating pumps, and other mechanical and electrical equipment; engine room.

Boiling. Heated to the boiling point; the change of a liquid to a gas.

Boiling Point. The temperature at which a liquid boils; the point at which a liquid starts to change to gas; boiling temperature of a liquid under a pressure of 14.7 psia (760 mm); water boils at 100° C or 212° F.

Boiling Temperature. Boiling point; temperature at which a fluid changes from a liquid to a gas.

Bole. 1. The main stem of a tree of substantial diameter; roughly, capable of yielding saw timber, veneer logs, or large pole; seedlings, saplings, and small diameter trees have stems, not boles. 2. A fine soft clay, yellow or dark, colored by iron oxide, formerly used as a pigment.

Bollard. 1. Short steel post, usually filled with concrete, set to prevent vehicular access to or to protect property from damage by vehicular encroachment. 2. Steel or cast iron post to which ships are tied.

Bollard Pipe. Short pipe length, placed vertically in the ground and filled with concrete to prevent vehicular access or to protect property from damage by vehicular encroachment.

Bolster. 1. A short piece of timber set horizontally across the top of a post, either to afford a greater bearing surface for a girder or girders, or to allow a post above to set between the ends of the girders, or to shorten the span of girders. 2. A long wire type chair used to support steel reinforcing bars in a concrete slab while the concrete is being placed.

Bolster, Slab. Continuous, individual support used to hold steel reinforcing bars in the proper position.

Bolt. 1. A threaded metal rod or pin for joining parts, having a head and usually used with a threaded nut. 2. a sliding bar for locking a door or gate. 3. A bar in a lock, moved by a key.

Bolt, Anchor. Steel bolt used to secure wood construction to concrete or masonry construction.

Bolt, Carriage. A threaded bolt with a round smooth head and a square neck directly under the head to prevent rotation.

Bolt Cutter. A hand tool, utilizing effective leverage, that can shear bolts and steel reinforcing rods.

Bolt, Dead. A lock bolt having no spring action nor bevel, and which is operated by a key or a turn piece.

Bolt, Door. The tongue of a lock installed to prevent a door opening.

Bolted Steel. Steel structural system where the members are assembled and connected with bolts, as opposed to welding.

Bolted Truss. Beams, frames, trusses, or other supports connected to support a roof, bridge, or floor system, that are fastened together with bolts.

Bolting Pattern. The arrangement, spacings, and dimensions of bolts used to attach two or more structural members together.

Bolt, Wood. Bolts specifically used in wood applications, that have an unslotted oval head and square shoulders that sink into the wood to prevent turning.

Bond. 1. The solid connection of one material to another; a substance which causes such a joining to take place. 2. The adhesion between masonry units and mortar or grout. 3. The patterns and methods in which brick and block are installed, for example, American bond, basket weave, Dutch cross bond, Flemish bond, running bond, and stack bond. 4. The adhesion between the surface of a reinforcing bar and the adjacent concrete, mortar, or grout. 5. The adhesion of cement paste to aggregate. 6. The degree of firmness with which the paper adheres to the gypsum board core. 7. The junction of the weld metal and the base metal. 8. The adherence of the bitumens between two layers of roofing felts. 9. See Chemical Bond. 10. See Mechanical Bond. 11. See Surety Bond. 12. See Completion Bond. 13. See Performance Bond. 14. See Payment Bond. 15. See Roof Bond. 16. See Bid Bond.

Bondability. 1. Indicating ease or difficulty in bonding a material with adhesive. 2. Ease or difficulty of a contractor in obtaining a surety bond.

Bond Beam Block. A concrete masonry unit with the upper part of the ends and webs removed to make room for horizontal reinforcing bars and grout. U-blocks are sometimes used to form bond beams, especially as over openings.

Bond Beam. A horizontal grouted element within a masonry wall in which steel reinforcement is embedded; a horizontal reinforced masonry beam, serving as an integral part of the wall.

Bond Breaker. A material used to prevent adhesion of newly placed concrete to other surfaces.

Bond Coat. A material used between the back of the tile and the prepared surface; suitable bond coats include pure portland cement. dry-set portland cement mortar, latex-type portland cement mortar, organic adhesive, and the like.

Bond Course. The course consisting of units which overlap more than one wythe of masonry.

Bonded Rubber Cushioning. Rubber or latex cushioning adhered to the carpet at the mill.

Bonded Stop Notice. A stop notice that is accompanied by a surety bond which guarantees any costs assessed against the claimant should the claimant lose its lawsuit; see Stop Notice.

Bonded Tendon. A prestressed tendon that is bonded to concrete either directly or through grouting.

Bonded Wall. A masonry wall in which two or more wythes are bonded to act as a structural unit; see Bonder.

Bonder. A bonding unit; see Header, 7.

Bonderized Flagpole. A metal flagpole that has been coated with a patented phosphate solution for protection against the elements.

Bonderizing. A five-step proprietary custom process for phosphatizing.

Bonding. Adhesion.

Bonding Agent. A substance applied to a suitable substrate to create a bond between it and a succeeding layer as between a subsurface and a terrazzo topping or a succeeding plaster application.

Bonding Jumper. 1. A reliable conductor to assure the required electrical conductivity between metal parts required to be electrically connected. 2. At the main service, the connection between the grounded circuit conductor and the equipment grounding conductor. 3. At equipment, the connection between two or more portions of the equipment grounding conductor. 4. In a circuit, the connection between portions of a conductor in a circuit to maintain required ampacity of the circuit.

Bond Plaster. A specially formulated gypsum plaster designed as a first coat application over monolithic concrete.

Bone Ash. Calcined bone consisting essentially of calcium phosphate.

Bone Black. Pigment made from calcined animal bones; dark in color, but does not have a strong tinting strength like lampblack.

Bone China. A translucent china made from a ceramic whiteware body composition containing a minimum of 25 percent bone ash.

Bonnet. The upper portion of the gate valve body into which the disc of a gate valve rises when it is opened.

Bonus. An extra payment to a contractor for achieving some specified goal, such as early completion.

Bookkeeper. A person who records the accounts or transactions of a business.

Book Matched. Wood veneer where adjoining sheets are sliced from the same log so that the grain matches like an open book.

Boom. 1. Along beam extending out from an upright to lift or carry something and guide it as needed; a derrick boom. 2. A barrier across a river or around an area of water to prevent floating logs from dispersing or to contain an oil spill. 3. To undergo swift, vigorous growth or development; flourish. 4. A period of business prosperity, or industrial expansion. 5. A sudden favorable turn in business prospects.

Booster. Common term applied to the use of a compressor when used as the first stage in the cascade refrigerating system.

Booth. 1.A small temporary roofed or unroofed structure used as a market stall. 2. A small enclosure used for various purposes, as for telephoning or voting. 3. A set of a table and benches in a restaurant or bar.

Booth, Spray. An area in a building or structure used for spray painting; blocked off by walls to prevent dust and dirt from work surface.

Border. The edge or boundary of anything; verge.

Bore. 1. To drill a hole. 2. The diameter of a tube. 3. Inside diameter of a motor cylinder. 4. A high tidal wave rushing up a narrow estuary.

Bored Lock. A door lock manufactured for installation in a circular hole.

Boring. 1. Making holes in wood or metal to aid in the insertion of bolts, nails or other fasteners. 2. Drilling into the ground to bring up samples of earth for testing. 3. Rotary drilling.

Borrow. Excavated material that has been taken from one area to be used as fill at another location.

Borrowed Light Opening. A glazed window unit in an interior partition.

Borrow Pit. An area designated as the source of earth removal to be used elsewhere for fill.

Boss. 1. A round knob, stud, or other protuberance on the center of a shield or in ornamental work. 2. A piece of ornamental carving covering the point where the ribs in a vault or ceiling cross. 3. The employer or contractor on a construction job.

Boston Ridge. Applying asphalt or wood shingles at the ridge or at the hips of a roof as a finish.

Bosun's Chair. See Boatswain's Chair.

Bottle Cooler. A container used for cooling or maintaining the coolness of bottled liquids.

Bottom Bars. The reinforcing bars that lie close to the bottom of a reinforced concrete beam or slab.

Bottom Beam. The lowest horizontal member supporting a building or structure.

Bottom Chord. The bottom member in a truss.

Bottom Dip. The lowest water or waste point in a trap.

Bottom Line. 1. The last line on a financial report which shows the profit or loss. 2. The most important factor or statement.

Bottom Pivot Hinge. A flexible pair of plates joined by a pin to allow swinging of a door or gate installed at the bottom.

Bottom Plate. In wood stud framing construction, the bottom continuous horizontal member that supports the studs.

Boulder. A large stone worn smooth by erosion; a rock which is too heavy to be lifted readily by hand.

Bounce-Back. Spray rebound similar to blow-back.

Boundaries. A separating line that indicates or fixes a limit or extent.

Boundary Elements. Portions along wall and diaphragm edges strengthened by longitudinal and transverse reinforcement; boundary elements do not necessarily require an increase in the thickness of the wall or diaphragm.

Boundary Survey. A survey of the property lines of a piece of land.

Bourdon Tube. Thin-walled tube of elastic metal flattened and bent into circular shape which tends to straighten as pressure inside is increased; used in pressure gauges.

Boutique. A small fashionable store or a small specialty shop within a larger store.

Bow. The distortion of lumber in which there is a deviation, in a direction perpendicular to the flat face, from a straight line from end-to-end of the piece.

Bowl, Toilet. The oval part of a toilet which receives the waste and fills with water after flushing the toilet tank.

Bow Trowel. A finishing trowel with a slight curve for crowning the final application of gypsum board joint treatment.

Bow Window. A bay window with a curved front.

Box Beam. A beam of metal, concrete, or plywood which, in cross section, resembles a closed rectangular box.

Box Culvert Formwork. The temporary wooden structure which holds wet concrete in place for the final curing of a rectangular-shaped, reinforced concrete drainage system.

Box Culvert Reinforcing. Iron or steel rods that are embedded in the wet concrete of a rectangular-shaped drainage system to give additional strength.

Box Culvert. A concrete draïnage structure rectangular shaped, reinforced and cast in place or made of precast sections.

Box Cutter. A specially designed hand tool for shear-cutting electrical outlet holes in gypsum board.

Box, Distribution. A box which contains the circuit breakers, connects to the service wires, and delivers current to the various outlets throughout a building or structure.

Box, Floor. A metal electrical rough-in box fed by conduits in or under the floor to provide for a floor outlet.

Box, Gang. Electrical rough-in box constructed of metal or hard plastic, to provide for two or more outlets or switches.

Boxing. Mixing by pouring back and forth from one container to another.

Box, Junction. A metal or hard plastic electrical rough-in box usually square or octagonal housing only wire or cable connections.

Box, Plastic. An electrical box for the joining of electrical wires, constructed of hard plastic and nailed in place.

Box, Pull. An electrical rough-in box placed in a length of conduit, through which cables can be pulled.

Box, Screw Cover. A removable ornamental or protective plate that is mounted to an electrical rough-in box.

Box System. A framing system that is made up of bearing walls with lateral forces being resisted by shear walls and diaphragms.

Box, Tap. The electrical box where the public service electrical supply line is connected with a branch to serve a particular building or structure.

Box, Terminal. A metal electrical box, usually with a removable cover, that contains leads from electrical equipment ready for connection to a power source.

Box, Weatherproof. An electrical box, designed for exterior installation, that is impervious to the outside elements, such as water.

Box Wrench. A type of end wrench in which the gripping end surrounds the nut or bolt head.

Boyle's Law. A physical law governing the behavior of gases, stating that the volume of a gas is directly proportional to its temperature and inversely proportional to its pressure.

Brace. 1.A cranked hand tool to hold a bit for drilling holes. 2. A diagonal member, either in tension or compression, to strengthen a structure.

Braced Frame. A truss system or its equivalent which resists lateral forces.

Braced Frame. One which is dependent upon diagonal braces for stability and capability to resist lateral forces.

Braced Wall Line. A series of braced wall panels in a single story.

Braces. Pieces fitted and firmly fastened to two others at any angle in order to strengthen the structure.

Bracing. 1. Diagonal members, either temporary or permanent, installed to stabilize a structure against lateral loads. 2. Structural member used to prevent buckling or rotation of wood studs.

Bracket. 1. A projecting support for a shelf or other structure. 2. In furring and lathing, a superficial structure usually in angles forming a frame to support lath; used to save material and weight in ornaments or cornices.

Bracket Hanger. Hanger supporting a wall-hung sink.

Bracket, Wall. 1. A wall-mounted support for shelving or other object. 2. A wall-mounted lighting fixture.

Brad. A small slender wire nail with a thickened top for a head.

Brake. A machine for flanging, bending, or folding sheet metal.

Brake Metal. Sheet metal that has been bent into a specified configuration, such as gravel stop, flashing, L-shapes, and Z-shapes; formed on a sheet metal brake.

Branch. A member or part of a system or structure which diverges from the main portion, as in heating, ventilation, or electrical installations; a smaller or subordinate duct, pipe, or circuit extending from the main line.

Branch Breaker. A switch which stops the flow of current by opening the circuit automatically when more electricity flows through the circuit than the circuit is capable of carrying; resetting may be either automatic or manual.

Branch Circuit. That portion of a wiring system extending beyond the final automatic overcurrent protective device, excluding any thermal cutout or motor running overload protective device that is not approved for short circuit protection.

Branch Circuit (General Purpose). A branch circuit that supplies a number of outlets for both lighting and appliances.

Branch Circuit (LIGHTING). Circuits supplying energy to lighting outlets only.

Branch Circuit (MOTOR). Circuits from the motor branch circuit protective device to the motor, including the controller and overload protective device.

Branch Circuit (MULTIWIRE). A multiwire branch circuit is a circuit consisting of two or more ungrounded conductors having a potential difference between them, and an identified grounded conductor having equal potential difference between it and each ungrounded conductor of the circuit and which is connected to the neutral conductor of the system.

Branch Circuit Appliance. Circuits supplying energy either to permanently wired appliances or to attachment plug receptacle, that is appliance or convenience outlets, or to a combination of permanently wired appliances and additional attachment plug outlets on the same circuit; such circuits to have no permanently connected lighting fixtures not a part of an appliance.

Branch Intervel. This is the vertical distance, generally a floor or story in height, but never less than 8 feet, within which the horizontal branches from one floor of the building are connected to the main DWV stack.

Branch Vent. A vent pipe that connects a branch of the drainage system to the main stack.

Brass. A metal alloy consisting essentially of copper and zinc in variable proportions.

Brass Fitting. Threaded pipe connector made of brass, used to join two pieces of pipe together.

Braze Welding. Sometimes known as Bronze Welding. Often carried out as in fusion welding except that the base metal is not melted. The base metal is simply brought up to what is known as a tinning temperature (dull red color) and a bead deposited over the seam with a bronze filler rod.

Braze. Solder with an alloy of brass and zinc at a high temperature.

Brazed Connection. Parts that are hardened and connected by soldering with an alloy.

Brazing. A welding process wherein coalescence is produced by heating to suitable temperatures above 800° F. and by using a nonferrous filler metal having a melting point below that of the base metals; the filler metal is distributed between the closely fitting surfaces of the joint by capillary attraction.

Breach of Contract. A material failure to perform an act required by contract.

Breadboard. A draw board in a kitchen cabinet used for kneading and slicing bread.

Break. 1.To damage by separating into pieces under a blow or strain; shatter. 2. An interruption as in a circuit. 3. An interruption in the continuity of an element, such as a plastered wall or cornice.

Breakdown. An itemized list of building costs.

Break Joints. To arrange joints so that they do not come directly under or over the joints of adjoining pieces, as in shingling, siding, and brick laying.

Breaker, Circuit. A switch which stops the flow of current by opening the circuit automatically when more electricity flows through the circuit than the circuit is capable of carrying; resetting may be either automatic or manual.

Breaker, Main. A switch in a main electrical service panel where the service wires attach.

Breaker Strip. Strip of wood or plastic used to cover joint between outside case and inside liner of refrigerator.

Breaker, Vacuum. An electrical breaker with a space that contains reduced air pressure.

Break-Even Point. The financial position wherein total revenue received equals the sum of the costs and expenses for a particular project showing neither profit nor loss.

Breaking Joints. The laying of bricks so that no two vertical joints come directly over one another; this is done for strength.

Breakpoint. The point at which a rising concentration of chlorine in swimming pools kills germs and bacteria by oxidizing organic matter; once all matter is oxidized, the amount of chlorine remaining is free or uncombined.

Breast. 1.The front part of a fireplace above and around the firebox opening. 2. The projecting portion of a chimney, especially when projecting into a room.

Breast Drill. A portable drill equipped with a plate that the operator leans against to provide pressure.

Breastsummer. A beam to carry the load above a fireplace opening, also called bressumer.

Breastplate. The pressure plate of a breast drill.

Breathing Zone. Area of a room in which occupants breathe as they stand, sit, or lie down.

Breeching. Space in hot water or steam boilers between the end of the tubing and the jacket.

Bressumer. See Breastsummer.

BRI. Building-Related Illness.

Brick. A solid masonry unit having the shape of a rectangular prism; usually made from clay, shale, fire clay, or a mixture of these.

Brick Anchor. Fasteners that are designed to attach and secure a brick veneer to a concrete or brick wall.

Brick and Brick. The laying of bricks by which the bricks are touching each other and the mortar used is just enough to fill the irregularities of the bricks.

Brick Bat. Part of a brick, usually half a brick or less.

Brick Bond. The pattern or arrangement of bricks in a wall.

Brick, Chimney. Brick, chosen for the specific use in the construction of chimneys, because of its ability to withstand high temperatures without cracking.

Brick, Fire. Brick that has been tested for fire-resistance and then graded for specific construction uses; brick whose composition and characteristics make it suitable to use for masonry fireplace fire-boxes and fire chambers.

Brick Firewall. A masonry wall constructed to prevent or retard the spread of fire.

Bricklayer. A brick mason.

Brick Manhole. A vertical access shaft from the surface to an underground area, constructed of bricks.

Brick Mason. A mason who builds in brick; also called a brick layer; see Mason.

Brick, Masonry. Bricks that are shaped and molded in different sizes and shapes.

Brick Molding. Milled trim piece designed to fill the gap between finished milled parts like door and window frames and irregular surfaces like masonry.

Brick, Paver. Brick units that are used in foot traffic areas; usually four inches wide, eight inches long, and 1-5/8 to 2-1/4 inches thick.

Brick Removal. The act or process of the demolition of a brick structure.

Brick Tongs. An iron grasping device consisting of two bars pivoted in the middle like a scissors and allowing two workers to lift and move a small pile of bricks.

Brick Trowel. The brick trowel is larger than the buttering trowel; the primary tool of masons; used when any preparatory brick work has to be done; its greater surface and weight are advantageous in the buttering and tapping in of block, brick, or larger tiles.

Brick Veneer. A one brick thick outside facing of brickwork used to cover a wall of some other material.

Brickwork. Masonry of bricks and mortar.

Bridge. A straightedge used as a starting line for the laying of tile; it can be blocked up to support tile over an opening.

Bridge Crane. A hoisting device spanning two overhead rails; the hoisting device moves laterally along the bridge with the bridge moving longitudinally along the rails.

Bridge Deck. The slab or other structure forming the travel surface of a bridge.

Bridge Glaze. Colorless or colored ceramic glaze having high gloss.

Bridging. 1. Diagonal or longitudinal members used to keep horizontal members properly spaced, in lateral position, vertically plumb, and to distribute load; pieces fitted in pairs from the bottom of one floor joist to the top of adjacent joists, and crossed to distribute the floor load; sometimes pieces of width equal to the joists and fitted neatly between them. 2. In painting, forming a skin over a depression.

Bridging Architect. An owner's architect who designs the project and then is replaced by the design/build entity's architect who prepares the construction documents.

Briding. A section sized to fit inside the flanges of studs and channels to stiffen construction.

Bright Blast. White blast.

Bright Glaze. A high-gloss coating with or without color.

Brilliant Color. Very bright

Brine. Water saturated or strongly impregnated with salt.

Brinell Hardness Test. A laboratory test for measuring the hardness of a material by hydraulically pressing a steel ball into the surface.

British Thermal Unit (BTU). Quantity of heat required to raise temperature of one pound of water one degree Fahrenheit.

Brittle. Easily broken; not tough.

Brittle Failure. Failure in material which generally has a very limited plastic range; material subject to sudden failure without warning.

Broach. 1. Any of various pointed or tapered tools, implements, or parts. 2. A drill bit. 3. A pointed tool for roughly dressing stone.

Broad Knife. A wide flexible finishing knife for applying joint finishing compound.

Broadcast. To sprinkle solid particles on a surface.

Broad-Leaved Trees. See Hardwoods.

Broadloom. Carpet woven on a broad loom in widths of 6 feet or more. usually 6, 9, 12, 15 and 18-ft. widths, and up to 30 ft.

in Chenille; broadloom is not a type of weave of carpet, nor a pattern or color; it is simply a designation of width.

Broiler, Kitchen. A cooking device in a kitchen that cooks food by direct exposure to radiant heat.

Broker. An agent who, for a fee or commission, brings parties together for the leasing or sale of real property.

Bromide. A chemical compound containing bromine, a halogen; sodium or potassium bromide in solution will produce free bromine if chlorine is added to the pool.

Bronze. An alloy of copper and tin and sometimes other elements.

Bronze Tools. Non-sparking tools; used when fire hazards are particularly acute.

Bronze Welding. See Braze Welding.

Bronzing. Formation of metallic sheen on a paint film.

Bronzing Liquid. A vehicle especially formulated for use as a binder for aluminum, gold, or bronze powder.

Broom. 1. A long handled brush of bristles for sweeping and brushing; also used for concrete finishing. 2. To spread out in a broom shape through separation of the fibers, as when a pile is partly crushed at its head under the blows of the pile driver.

Broom Finish. A finish applied to an uncured concrete surface, to provide skid or slip resistance, made by dragging a broom across the freshly placed concrete surface.

Brooming. In roofing, embedding a ply by using a broom to smooth out the ply and ensure contact with the adhesive under the ply.

Brown Coat. The coat of plaster directly beneath the finish coat; in two-coat work, brown coat refers to the basecoat plaster applied over the lath; in three coat work, the brown coat refers to the second coat applied over a scratch coat; brown coats are applied with a fairly rough surface to receive the finish coat.

Brownfield Site. A building site that has been previously built upon and is possibly polluted with toxic chemicals; compare with Greenfield Site.

Brown Out. To complete application of basecoat plastering.

Brown Stain. Brown discoloration of the sapwood of some pines caused by a fungus.

Brownstone. 1. A building sandstone of prevailing brown color. 2. A dwelling faced with brownstone.

Brush. 1. A painter's implement composed of bristles set into a handle, used for applying paint and other coatings to a surface. 2. An electrical conductor that makes contact with a moving part.

Brushability. 1. Adaptability of paint to application with a brush. 2. The ability or ease with which a paint can be brushed.

Brushcoating. The application of paint, stucco color, or other materials with a broad brush.

Brush Cutting. The act of removing unwanted plants to clear an area.

Brushed Surface. A sandy texture obtained by brushing the surface of freshly placed or slightly hardened concrete with a stiff brush for architectural effect or, in pavements, to increase skid resistance.

Brush Hand. A painter whose ability lies in his skill in applying material.

Brush Hook. A long handled tool for clearing brush.

Brush-Off Blast. Lowest blast cleaning standard.

Brussels. A term formerly, but now rarely, used to describe a loop pile or round-wire carpet woven on the Wilton loom.

Brutalism. A heavy plain style of architecture.

BTU. British thermal unit; measurement of the heat energy required to raise one pound of water one degree Fahrenheit.

Bubble. A large void in the core of gypsum board caused by the entrapment of air while the core is in a fluid state during the manufacturing process.

Bubbling. Describes the appearance of bubbles on the surface while a coating is being applied.

Buck. Assembly of the framing that constitutes a rough door or window opening.

Bucket. Large metal container into which concrete is discharged; the bucket is raised by crane to the placement area.

Bucket Trap. A mechanical steam trap operating on buoyancy that prevents the passage of steam through the mechanical system it protects.

Buckle. To bend under compression; with very thin members, the bucking may be elastic, and the member will spring back if the load is removed; if the load is continued or if the buckling occurs with the stresses above the yield point, the member will fail by collapsing completely.

Buckles. In old plastering, raised or ruptured spots which eventually crack, exposing the lath beneath; most common cause for buckling is application of plaster over dry, broken or incorrectly applied wood lath.

Buckling. 1. Collapse, in the form of a sudden sideways deflection, of a slender element subjected to compression; structural failure by gross lateral deflection of a slender element under compressive stress, such as the sideward buckling of a long, slender column or the upper edge of a long, thin floor joist. 2. Wrinkling or ridging of the carpet after installation, caused by insufficient stretching, dimensional instability, or manufacturing defects.

Budget. A managerial plan of proposed operations to accomplish a financial objective.

Buffer. A pile of blasted rock left against or near a face to improve fragmentation and reduce scattering from the next blast.

Buffing. A router trimming of the shear cut end of gypsum board to smooth cut and adjust for length tolerances prior to the bundling tapes being applied.

Buffing Compound. Soft abrasive in stick form, bonded with wax.

Buglehead Screw. A special design screw that will seat beneath the plane of the gypsumboard surface without tearing the paper.

Build. Construct by putting materials or parts together.

Builder. One who constructs; a building contractor.

Builders' Hardware. Hardware used in construction, such as bolts, nuts, screws, nails, and other fastenings, metal and plastic parts, hinges, butts, catches, and similar parts.

Builder's Level. A simple form of transit-level for measuring and setting levels on a construction site; also called a contractor's level.

Building. A fixed structure forming an enclosure and providing shelter.

Building Brick. Brick for building purposes not especially treated for texture or color; also called Common Brick.

Building Code. A set of rules governing the quality of construction in a community; the purpose of these rules is to protect the public health and safety.

Building Coverage. The percentage of a specific parcel of land covered by buildings.

Building Demolition. The destruction, by means of explosives or otherwise, of a standing construction.

Building Drain. The lowest part of a drainage system which receives the discharge from soil, waste and other drainage pipes inside the walls of the building and conveys it to the building sewer beginning 3 feet outside the building wall.

Building Drainage System Complete system of piping used for carrying away waste water and sewage; also called House Drainage System.

Building Envelope. The outer limits of a building as encompassed by the exterior walls, roof, and foundation.

Building Excavation. The act or process of removing an area of earth and rock to make room for a foundation.

Building Inspector. A public official that examines the work in the field to determine compliance with the applicable building laws.

Building Integrated Photovoltaic Cell. Photovoltaic cells that substitute for building elements or materials, such as spandrel panels or roofing panels.

Building Lime. A lime whose chemical and physical characteristics and method of processing make it suitable for ordinary or special construction uses; also called construction lime.

Building Main. Water supply pipe that carries the water from the source of supply to the first branch of the water distributing system in the building.

Building, Metal. A building or structure constructed of a structural steel frame covered by metal roof and wall panels; commonly prefabricated in a factory and assembled at the site.

Building Official. The official charged with administration and enforcement of the applicable building code, or his duly authorized representative.

Building Paper. Any of various usually water repellent sheet materials such as kraft paper assemblies and asphalt saturated felt, used for moisture proofing, draft stopping, and exterior plaster backing; a general term for papers, felts, and similar sheet materials used in buildings, without reference to their properties or uses.

Building Related Illness. Diagnosable illness whose symptoms can be identified and whose cause can be directly attributed to airborne building pollutants, for example, Legionnaire's disease, and hypersensitivity pneumonitis.

Building Residual. A real estate appraisal technique where a reasonable return on the land is first deducted from the income, the balance being attributable to the improvements.

Building Sewer. The piping that takes the soil and waste water from the building drain and conveys it to the public sewer or private sewage disposal system; also called House Sewer.

Building Sub-Drain. That part of the drainage system of a house that receives the discharge from the fixtures and cannot drain into the building sewer because it is located below the building sewer or building drain; this piping usually drains into a sump and is pumped up into the house sewer or building drain.

Building Trap. A trap placed in the building drain to prevent entry of sewer gases from the sewer main.

Build-Up Sequence. The order in which the weld beads of a multiple-pass weld are deposited with respect to the cross-section of the joint.

Built-In Oven. An oven installed directly into a wall or cabinet.

Built-In Shelving. Shelving that is permanently installed in a cabinet frame.

Built-Up Member. A single structural component made from several pieces fastened together.

Built-Up Plaque. Layers of a localized abnormal patch on a surface.

Built-Up Roof Removal. The act or process of removing many layers of old roofing material.

Built-Up Roof. A roof covering made of continuous rolls or sheets of saturated or coated felt, cemented together with bitumen, and may have a final coating of gravel or slag.

Built-Up Roofing Membrane (BUR). Roofing material applied in sealed waterproof layers or plies, or saturated or coated felts alternated with layers of bitumen, surfaced with mineral aggregate or asphaltic materials.

Built-Up Steel Lintel. Lintel fabricated of two or more pieces of structural steel secured together to act as one member.

Built-Up Timber. A timber made of several pieces fastened together and forming one of larger dimension.

Bulb, Sensitive. Part of sealed fluid device which reacts to temperature; used to measure temperature or to control a mechanism.

Bulb Tee. Rolled steel in the form of a T with a formed bulb on the edge of the web.

Bulge. An irregular swelling; a lump.

Bulk. 1. Size; magnitude; large mass or quantity. 2. Large quantities as in bulk cement. 3. To cause to swell or bulge. 4. A small structure projecting from a building, as a booth or stall.

Bulk Density. The weight of a material per unit of volume.

Bulked Continuous Filament (BCF). Continuous strands of synthetic fiber made into yarn without spinning; often extruded in modified cross section such as multi-lobal, mushroom or bean shape and/or texturized to increase bulk and covering power.

Bulk Excavation. The digging out of large amounts of dirt and debris.

Bulkhead. 1.A vertical partition separating compartments, as on a ship or aircraft. 2. A structure or partition to resist pressure or shut off water, fire, or gas. 3. A retaining wall along a waterfront. 4. A projecting structure with a sloping door giving access to a cellar stairway or shaft.

Bulkhead Formwork. The temporary formwork that blocks fresh concrete from a section of forms or closes the end of a form at a construction joint.

Bulking. Increase in the bulk volume of a quantity of sand in a moist condition over the volume of the same quantity dry.

Bulking Curve. Graph of change in volume of a quantity of sand due to change in moisture content.

Bulking Factor. Ratio of the volume of moist sand to the volume of the sand when dry.

Bulking Value. Of a pigment, its ability to add volume to a paint.

Bulk Mail Slot. An opening in a wall or door big enough to receive large pieces of mail.

Bulldozer. A tractor driven machine with a horizontal blade for clearing land, road building, or similar work.

Bulletin Board. A wall hanging unit upon which information or messages are attached; a thin board, often of cork, hung on a wall for the attachment of public announcements.

Bulletproof. Manufactured to prevent breakage or penetration from a strong external force.

Bulletproof Glass. A protective laminated glass sheet, heat and pressure-bonded with resin sheets to resist the passage of bullets.

Bulletproof Partition. A dividing wall which has been rendered bulletproof by using specific materials in its construction.

Bull Float. A tool comprising a large, flat, rectangular piece of wood, aluminum, or magnesium usually 8 in. (20 cm) wide and 42 to 60 in. (100 to 150 cm) long, and a handle 4 to 16 ft. (1 to 5 m) in length used to smooth unformed surfaces of freshly placed concrete.

Bull Nose Brick. 1. A brick having one rounded corner. 2. Bricks that have their ends or corners rounded off.

Bull Nose. 1. In plastering, an external angle that is rounded in order to eliminate a sharp corner; can be made by running with plaster or using a bull nose corner bead with the proper radius. 2. Any material with a rounded edge such as a concrete block, ceramic tile, or brick. 3. See Step Return.

Bullnose Block. A concrete masonry unit which has one or more rounded external corners.

Bullnose, Corner. A type of ceramic tile bullnose trim with a convex radius on two adjacent edges.

Bullnose, Glazed. A ceramic trim tile with a convex radius on one edge that has been given a glassy or glossy surface.

Bull's Eye. 1. A small circular window or opening; a bull's eye window. 2. A small thick disc of glass inserted, as in a deck, to let in light.

Bumper, Dock. Thick rubber units placed under loading dock openings to absorb the shock and prevent damage when trucks back in for loading or unloading.

Bumper, Door. Rubber tip devices mounted on walls or baseboards that prevent door knobs from marring walls.

Bundle. Two pieces of gypsumboard packaged face to face.

Bundled Bars. A group of not more than four parallel steel reinforcing bars in contact with each other, usually tied together.

Bundling Tape. End tape to secure two pieces of gypsum board into a bundle.

Bund Wall. A wall built around a tank to contain its contents should the tank or its piping spring a leak.

Bunker. Space where ice or cooling element is placed in commercial installations.

Buoyant Uplift. The force of water or liquefied soil that tends to raise a building foundation out of the ground.

BUR. Built-Up Roofing.

Burden. The amount of money that has to be added to cover overhead.

Burglar Alarm. A security system that signals when any of the contacts have been interrupted.

Burglar Bars. A grille of steel bars to protect a window or skylight opening from intrusion.

Burl. A hard woody outgrowth on a tree, good for highly figured veneers.

Burlap. A coarse fabric of jute, hemp, or less commonly, flax, for use as a water-retaining covering in curing concrete surfaces; also called Hessian.

Burlap, Concrete. A curing concrete surface that has had a coarse fabric of jute, hemp, or less commonly, flax applied, for use as a water-retaining covering.

Burlap Rub. A finish obtained by rubbing burlap to remove surface irregularities from concrete.

Burling. In carpet manufacturing, a hand-tailoring operation after weaving, to remove any knots and loose ends, to insert missing tufts of surface yarn and otherwise check the condition of the fabric; also, a repair operation on worn or damaged carpet.

Burn. To cut metal with a gas flame.

Burned. Over-dried, partially calcined gypsum board.

Burned Finish. Wood finish in which hard portion of grain stands out in relief; produced by using blowtorch and stiff bristled brush.

Burner. Device in which burning of fuel takes place.

Burning In. Repairing a finish by melting stick shellac into the damaged places by using a heated knife blade or iron.

Burnish. Polish by rubbing.

Burnoff Rate. See Melting Rate.

Burnt Sienna. An earthy substance containing oxides of iron and usually of manganese; orange red or reddish brown pigment, used in paint; sienna that has been roasted.

Burn-Thru. A term erroneously used to denote excessive Melt-Thru or a hole.

Burnt Umber. A brown earthy substance containing oxides of iron and manganese; a pigment, darker than ochre and sienna, used in paint; umber that has been roasted.

Burr. A sharp, roughened, in-turned edge on a piece of pipe which has been cut but not reamed.

Burst Strength. The internal pressure required to break a pipe or fitting; this pressure will vary with the rate of build-up of the pressure and the time during which the pressure is held.

Bus Bar. 1. A large, flat conductor, usually solid copper, used for carrying very high electrical currents. 2. An uninsulated bar or tube used as an electrical conductor at a circuit junction.

Bus Duct. A prefabricated unit containing one or more electric conductors, often a metal bar, that serves as a common connection for two or more circuits.

Bus Duct Connection. A metal bar serving as a common connection for two or more circuits in a prefabricated unit.

Bus Ground. In the main electrical service panel, where the neutral service wire, generally white, attaches and is linked to the earth by the ground wire.

Bush Hammer. In stone dressing, a steel hammer used in finishing the harder stones; it has a square-ended prismatic head divided into a number of pyramidal points.

Bushed Nipple. A pipe threaded at both ends to connect two pipes of different dimensions.

Bushing. 1. A removable cylindrical lining for an opening used to limit the size of an opening, resist abrasion, or serve as a guide. 2. An electrically insulating lining for a hole to protect a through conductor. 3. A pipe fitting with both male and female threads used in a fitting to reduce the size; used to connect pipes of different sizes.

Bushing, Conduit. A threaded metal or plastic pipe connector used to connect conduit to a box or other housing where the hole is not threaded.

Business Entity Concept. The assumption that a business is separate and distinct from its owner's financial operations and holdings.

Busway. A rigid assembly consisting of one or more busbars.

Butane. A gaseous hydrocarbon (C_4H_{10}) of the alkane series used in liquefied form as fuel; also used as a low temperature application refrigerant.

Butler's Pantry. A service room between kitchen and dining room

Butt. 1. The larger of the two ends of a log. 2. See Butt Hinge. 3. Pile Butt.

Butterflies. Color imperfections on a lime putty finish wall which smear out under pressure of the trowel; caused by lime lumps not put through a screen and insufficient mixing of the gauging.

Butterfly Reinforcement. Strips of metal reinforcement placed diagonally over the plaster base at the corners of openings before plastering.

Butterfly Roof. A roof shape that is like an inverted gable, the rain gutter being in the middle instead of the ridge.

Butterfly Tie. See TIES.

Butterfly Valve. A valve constructed with a disc that rotates 90 degrees within the valve body.

Buttering. 1. Spreading mortar on a masonry unit before it is laid. 2. The spreading of a bond coat to the backs of ceramic tile just before the tile is placed. 3. See Surfacing.

Buttering Trowel. The blade of the buttering trowel is approximately 4-1/2" wide and 7" long; used in buttering rich mortar to masonry, a method commonly used in the eastern states.

Butt Hinge. A type of hinge designed for mortising into the edge of the door and into the rabbet of a door frame, consisting of two plates with a removable connecting pin; also called a Butt.

Butt Joint. 1. A plain square joint between two members. 2. In wallpaper, a joint made by trimming both selvedges and butting the edges together; this is used in highest type of work. 3. The cut ends of gypsum board placed adjacent to one another. 4. The joint between two bricks placed end to end in the same course; also called Cross Joint, Head Joint, or Vertical Joint.

Buttonback Tile. Tile that have round or square projections on the bondable side.

Butt, Pile. See Pile Butt.

Buttress. 1. A projecting structure of masonry or wood for supporting or giving stability to a wall or building wall and to react against horizontal outward forces. 2. The broadened base of a tree trunk or a thickened vertical part of it.

Butt Weld. A weld in a butt joint between two members lying approximately in the same plane.

Butt Welded Pipe. Pipe that is joined by welding.

Butt-Welded Space. A reinforcing bar splice made by welding the butted ends.

Butyl Acetate. A lacquer solvent made from butyl alcohol by reaction with acetic acid.

Butyl Alcohol. An alcohol of higher boiling range than wood alcohol or grain alcohol; obtained from corn by fermentation.

Butyl Caulk. Caulking that is made from various synthetic rubbers derived from butanes.

Butyl Membrane. Pliable thin sheets or layers made from synthetic rubber.

Butylene Plastics. Plastics based on resins made by the polymerization of butene or copolymerization of butene with one or more unsaturated compounds, the butene being in greatest amount by weight.

Buzzer. An electric signaling device that produces a buzzing sound.

BX Cable. A type of indoor wiring consisting of two or more insulated wires protected by a wound, galvanized steel strip cover, the metal winding forming a flexible tube offering protection similar to rigid conduit.

BX Clamp. A clamping device to hold BX cable firmly in place against a wooden or metal member.

Bypass. 1. Passage at one side of, or around, a regular passage. 2. A secondary pipe or bridging of any plumbing fixture allowing this fixture to be disconnected and circulation maintained.

By-Product. A substance obtained during the manufacture of another substance; a secondary result, sometimes unexpected or unintended, of some process.

Byte. In computing, a group of eight binary digits.

C

C. 1. Celsius.2. Centigrade. 3. In heat transfer, conduction of a material to the passage of heat; the reciprocal of resistance (C=1/R). 4. In seismic design, the numerical coefficient used which represents building acceleration. 5. Capacitance.

Cabana. A shelter near a swimming pool or beach.

Cab, Elevator. The enclosure in an elevator which carries passengers and freight up or down.

Cabin. 1. A small one story dwelling of simple construction. 2. A compartment on a ship or aircraft.

Cabinet. 1. Case, box, or piece of furniture with sets of drawers or shelves, with doors, primarily used for storage. 2. An electrical switch enclosure designed either for surface or flush mounting, and provided with a frame, mat or trim in which swinging doors are hung; see Panelboard or Switchboard.

Cabinet, Base. Floor-mounted cabinet, usually with a counter, sink or appliance installed.

Cabinet Drawer Guide. A wood strip used to guide the drawer as it slides in and out of its opening.

Cabinet Drawer Kicker. Wood cabinet member placed immediately above and generally at the center of a drawer to prevent tilting down when pulled out.

Cabinet, Extinguisher. A case or cupboard having doors which contains a fire extinguishing device.

Cabinet Finish. Protective coatings to provide protection and decorative appearance for exposed portions of wood cabinets.

Cabinet Hardware. Metal and plastic fasteners and connectors used to facilitate the operation and movement of doors, drawers, and shelves in cabinets.

Cabinet Heater. A heating element enclosed in a metal housing, with openings for airflow, usually with a fan for controlling air flow.

Cabinet, Kitchen. A case, box, or piece of furniture with sets of drawers or shelves, with doors, primarily used for storage, mounted on walls or floors in a kitchen area.

Cabinet Knob. A handle, pull, or rounded protuberance for opening a cabinet door or drawer.

Cabinet, Laboratory. A case, box, or piece of furniture with sets of drawers or shelves, with doors, primarily used for storage, used in a place or area for scientific studies or commercial and institutional laboratories and testing facilities.

Cabinet, Monitor. A cabinet whose doors have louvered panels to allow for ventilation, light or finish design.

Cabinet, Transparent Finish. Protective clear or tinted coating to provide protection and decorative appearance for exposed portions of wood cabinets which allows grain of wood to be seen through coating.

Cabinet Wall. The wall to which cabinets are attached or mounted.

Cable. 1. A thin, flexible line which carries only tensile forces. 2. A bundle of two or more electrical conductors.

Cable, Audio. A cable which carries and transmits audio signals.

Cable Bus. An assembly of insulated cables.

Cable, BX. A type of indoor wiring consisting of two or more insulated wires protected by a wound, galvanized steel strip cover; the metal winding forms a flexible tube, offering protection similar to rigid conduit.

Cable Camera. A camera that is controlled and activated with coaxial cable.

Cable, Coaxial . A cable consisting of two concentric conductors separated by an insulator; usedto transmit telephone, television and computer signals.

Cable, Communication. A cable fortransmission of telephone, television, and computer signals.

Cable, Computer. Coaxial cable which transmits computer signals.

Cable Connector. A device used to connect lengths of cable together into one longer length.

Cable, Copper. Insulated, sheathed copper wires conducting power from a source to an electric appliance.

Cable Coupler. A device for connecting two lengths of cable into one longer length.

Cable, Fire Alarm. A specific electrical system cable which carries electric current to a warning horn or bell for use in the event of a fire or other catastrophe.

Cable Fitting. Couplings, elbows, tees or unions used to form a junction or connect cable lines together.

Cable, Guy. A wire used to secure a tall exterior mast, antenna, or other structure in place.

Cable Lug. A connector for fastening the ends of cable to a terminal.

Cable Man-Hole. A vertical access shaft from the surface to the underground, leading to an area for the repair or installation of cable wires.

Cable Receptacle. An interrupting outlet box device installed in an electric supply circuit for the connection of electric cables.

Cable Roof. A circular structure in which the internal stresses of the members are subjected primarily to tensile stresses.

Cable Support. A structure which holds cable lines in place or at a safe overhead height.

Cable Tap Box. A box where public cable service supply lines are connected with a branch to serve a building or structure.

Cable, Thermostat. A specific electrical system cable which operates an automatic device for regulating the temperature in a room, space, or area.

Cable Tray. Open track for support of insulated cables.

Cable TV. A system of distributing television signals to individual subscribers by use of subterranean cables or overhead wires, rather than by aerials.

CABO. Council of American Building Officials.

Cabriole Leg. A curved furniture leg rounded and swollen at the top and tapered down ending in an ornamental foot.

CAD. Computer Aided Design.

CADD. Computer Aided Design and Drafting.

Cadmium. A bluish white malleable ductile toxic metallic element used in protective platings and in bearing metals.

Cadmium Lithopone. A series of yellows and reds that are permanent to light and resistant to alkalis.

Cadmium Red. Non-fading red pigment made from cadmium and selenium metals; heat and alkali resistant.

Cadmium Yellow. Pigment prepared by precipitation from acid solution of soluble cadmium salt with hydrogen sulphide gas; fast to alkalis but not to acids.

Cafe. A small coffee shop or simple restaurant.

Cafeteria. A restaurant where the diners collect their food on a tray and usually pay before eating.

Cage Ladder. A, usually wall-mounted, ladder that has, for safety, a surrounding structure to prevent the climber from falling off.

Caging. Metal furring used to enclose pipes, columns, beams or other configurations to be concealed by gypsum board.

Caisson. 1. A type of drilled or augured piling. 2. A cylindrical, sitecast concrete foundation that penetrates through unsatisfactory soil to rest upon an underlying stratum of rock or satisfactory soil. 3. A foundation pier, either circular or rectilinear in plan, usually sunk to rock either by means of gravity, compressed air or by the open-well method. 4. A panel sunk below the normal surface in flat or vaulted ceilings.

Caisson Disease. A sometimes fatal disorder that afflicts workers in a compressed air atmosphere who return to normal air pressure too quickly; decompression sickness; also called The Bends, Air Embolism, or Aerembolism.

Caking. Hard settling of pigment from paint.

Calcareous. Containing calcium or calcium carbonate; chalky.

Calcification. Buildup of calcium carbonate on swimming pool walls and equipment; caused by precipitation of calcium from hard water.

Calcimine. A white or tinted wash consisting of glue, whiting or zinc white, and water, used primarily on plastered surfaces; a type of tempera; sometimes written Kalsomine.

Calcine. 1. To alter composition or physical state by heating. 2. To drive off or lose chemically combined water by action of heat thereby altering the chemical and physical characteristics of a material. 3. To release part or all of the water of crystallization from gypsum by the application of heat. 4. A ceramic mineral or mixture fired to less than fusion for use as a constituent in a ceramic composition.

Calcined. Heated to high temperature in absence of air.

Calcined Board. Gypsum board which has been subjected to excessive heat.

Calcined Gypsum. Gypsum that has been partially dehydrated by heat.

Calcium Aluminate Cement. The product obtained by pulverizing clinker consisting essentially of hydraulic calcium aluminates resulting from fusing or sintering a suitably proportioned mixture of aluminous and calcareous materials.

Calcium Carbonate. Earth product obtained from deposits of chalk or dolomite; also called Whiting; used as extender pigment.

Calcium Driers. Used widely in combination with other metal driers to convert paint to hard films.

Calcium Gypsum. A dry powder, primarily calcium sulfate hemihydrate, resulting from calcination of gypsum; cementitious base for production of most gypsum plasters; also called Plaster of Paris; sometimes called Stucco.

Calcium Hypchlorite. A chemical compound of chlorine and calcium used as a bacteriacide in swimming pools; available in white granular or tablet form and releases 70 percent of its weight as available chlorine.

Calcium Insulation. A type of insulation, made of hydrated calcium silicate, that can withstand 1200° Fahrenheit and is not affected by moisture.

Calcium Silicate. A sand and lime solution.

Calcium Sulfate. 1. The chemical compound $CaSO_4$. 2. White inert pigment which provides very little color or opacity. 3. A drying agent or desiccant in liquid line driers.

Calculus. A method of computation or calculation in a special notation; the use of algebra to calculate changing quantities.

Calendar Day. Each and every day on the calendar without deduction for weekends or holidays.

Calendered Papers. Wallpapers with hard finish.

Caliber. 1. The internal diameter of a tube. 2. The character and capacity of a firm.

Calibrate. To verify the graduations of an instrument and adjust them if necessary.

Calidarium. The room containing the warm bath in Roman baths, also called Thermae.

Caliper. 1.A measuring instrument with two legs resembling a divider, for measuring thicknesses, diameters, and distances between surfaces; with in-turned points to measure convex surfaces and out-turned points for measuring internal dimensions. 2. The precise measured thickness of gypsumboard. 3. The diameter of a tree trunk.

Call System. A communicating device to connect one place with another.

Calorie. 1. Small calorie or gram calorie, used by medical science;the amount of heat needed to raise the temperature of one gram of water one degree Celsius. 2. Large calorie or great calorie; used by engineering science; the amount of heat needed to raise the temperature of one kilogram of water one degree Celsius; a kilocalorie equals 1,000 calories.

Calorimeter. Device used to measure quantities of heat or determine specific heats.

Cam. A projection on a rotating part in machinery, shaped to impart reciprocal or variable motion to the part in contact with it.

Camber. 1.A deflection that is intentionally built into a structural element or form, usually a beam, to improve appearance or to nullify and offset the deflection of the beam under the effects of loads, shrinkage, and creep. 2. A slight rise at the center of a flat arch.

Camber Rod. A tension rod installed under a trussed beam.

Cambium. A thin layer of tissue in a tree between the bark and wood that repeatedly subdivides to form new wood and bark cells.

Came. Slender grooved lead rod used in stained glass windows.

Camera Cable. A device for the transmission or recording of visual images.

Campanile. A free-standing bell tower.

Campus. The grounds and buildings of a university, college, or school.

Candela. An international unit of luminous intensity; also called a Candle.

Candelabrum. (Pl. candelabra) A branched candle stick or lamp with several lights.

Candelilla Wax. Wax obtained from small shrub grown in Texas and Mexico; softer than Carnuba wax.

Candle. See Candela.

Candlepower. The unit of luminous intensity of a light source, expressed in Candelas.

Cane Bolt. An L-shaped rod, mounted on a swinging or sliding door or gate, that drops into a pipe sleeve below the floor surface to secure the door or gate.

Cannibalization. The process of operating an income property by borrowing parts, fixtures, and equipment from vacant spaces.

Cannular. Tubular.

Canopy. An overhanging shelter; a marquee.

Cantilever Wall. A retaining wall in which the wall and footing resists earth pressure by cantilever effect.

Cantilever. A structural shape, beam, truss, or slab, that extends beyond its last point of support.

Cantilevered Beam. A beam that is supported at one end only.

Cant Strip. 1. A strip of material, usually treated wood or fiber, with a sloping face used to ease the transition from a horizontal to a vertical surface at the edge of a flat roof; prevents the roofing material from abruptly stopping at the parapet wall and also helps prevent leakage at that juncture. 2. A triangular shaped strip of wood used under shingles at gable ends or under the edges of roofing on flat decks.

Cap. 1. A caplike part or thing; cover, or top. 2. An upper limit set on a budget or cost; a ceiling. 3. A trim tile with a convex radius on one edge; used for finishing the top of a wainscot or for turning an outside corner, a bullnose. 4. The upper member of a column, pilaster, door, cornice molding, and the like. 5. A female pipe fitting, solid at one end; used to close off the end of a piece of pipe. 6. Masonry units laid on top of finished masonry wall or pier. 7. Flashing for tops of parapet walls.

Capacitance Sensor. A sensor which shows a device's ability or hold or store electrical energy.

Capacitance. Property of a nonconductor, condenser or capacitor, that permits storage of electrical energy in an electrostatic field; a measure of this is the farad. 2. Property of an electric circuit which tends to oppose a change in voltage.

Capacitive Reactance. The opposition or resistance to an alternating current as a result of capacitance; expressed in ohms.

Capacitor. 1. A device which introduces capacitance into an electric circuit. 2. Type of electrical storage device used in starting and/or running circuits on many electric motors.

Capacitor Motor. Single-phase induction motor with an auxiliary starting winding connected in series with a condenser (capacitor) for better starting characteristics.

Capacitor-Start Motor. Motor which has a capacitor in the starting circuit.

Capacity. 1. The ability to contain or receive. 2. The maximum amount that can be accommodated or contained. 3.

Refrigeration rating system; usually measured in Btu per hour or watts (metric). 4. The requirements of a lender that indicate borrowers' physical and financial ability to proceed with a project and be able to withstand any reasonable loss.

Cap Base. Wood strip applied to the base of a wall to protect wall surface and finish the intersection of wall and baseboard.

Cap, Black Iron. A steel fitting, with female threads, which seals the end of a pipe.

Cap Block. A solid flat slab usually 2-1/4 inches thick used as capping units for parapet and garden walls; also used as a Paving Unit.

Cap Flashing. See Counterflashing.

Capillarity. The action by which the surface of a liquid where it is in contact with a solid, as in a capillary tube, is elevated or depressed depending on the relative attraction of the molecules of the liquid for each other and for those of the solid.

Capillary. A tube with a small bore.

Capillary Attraction. 1. The force of adhesion between a solid and a liquid in capillarity. 2. In brazing, the phenomenon by which adhesion between the molten filler metal and the base metals, together with surface tension of the molten filler metal, distributes the filler metal between the properly fitted surfaces of the joint to be brazed.

Capillary Break. A slot or groove intended to create an opening too large to be bridged by a drop of water, and thereby to eliminate the passage of water by capillary action.

Capillary Movement. Movement of underground water in response to capillary attraction.

Capillary Space. In cement paste, any space not occupied by anhydrous cement or cement gel; air bubbles, whether entrained or entrapped, are not considered as part of the cement paste.

Capillary Tube. See Capillary and Choke Tube.

Capillary Water. Underground water held above the water table by capillary attraction.

Capital. 1. The money or other assets with which a company starts in business and uses in operation. 2. The head or cornice of a pillar or column; also called Column Cap. 3. A city serving as the seat of government.

Capital, Column. The uppermost member of a column crowning the shaft and taking the weight of the slab, beam, or girder.

Capital Gain or Loss. The gain or loss in the sale or disposition of a capital asset.

Capitalization. 1.The process of determining the capital value of a real property investment by relating its annual income to an assumed capitalization rate; for example, a property with an annual income of $12,000 and a capitalization rate of 8 percent would have a capitalized value of $150,000 (Income/ Capitalization Rate = Capitalized Value). 2. Total of financial resources available to the business.

Capitalization Rate. The rate of interest to be used in the capitalization process, reflecting risk and rates of return on alternative investments; also called Cap Rate.

Capital Stock. The total amount invested in the business by the owner in exchange for shares of common stock at par value.

Capitol. 1.A building in which a state legislative body meets. 2. The building in which the US Congress meets in Washington, DC.

Cap, Pile. A structural member usually fastened to, and placed on the top of a slender timber, concrete, steel pile; used to transmit loads into the pile or group of piles and to interconnect them.

Cap, Post. A fitting which joins the end of a wooden post to a joist or girder connected to the post.

Cap Rate. See Capitalization Rate.

Cap Sheet. The top sheet of roofing in a built up roof, usually made of organic or inorganic fibers, saturated and coated on both sides with a bituminous compound and factory-coated with mineral granules, mica, talc, iliminite, asbestos, or other inorganic fibers, or similar material.

Cap, Welded. A fitting that is fastened by welding to seal the end of a pipe.

Car Wash. A building housing a mechanized method for washing cars.

Carbon. A nonmetallic element found native, as in the diamond and graphite, or as a constituent of coal, petroleum, and asphalt, of limestone and other carbonates, and of organic compounds; when combined with iron, forms various kinds of steel; in solid form, it is used as an electrode for arc welding; as a mold, it will hold weld metal; motor brushes are made from carbon.

Carbon-Arc Cutting. An arc-cutting process wherein the severing of metals is effected by melting with the heat of an arc between a carbon electrode and the base metal.

Carbon-Arc Welding. An arc-welding process wherein coalescence is produced by heating with an electric arc between a carbon electrode and the work and no shielding is used; pressure may or may not be used and filler metal may or may not be applied.

Carbonation. Reaction between the products of portland cement and carbon dioxide to produce calcium carbonate.

Carbon Black. Jet black, non-bleeding pigment, made by burning natural gas in insufficient supply of air.

Carbon Dioxide. Compound of carbon and oxygen (CO_2) which is sometimes used as a refrigerant; refrigerant number is R-744.

Carbon Filter. Air filter using activated carbon as air cleansing agent.

Carbonizing Flame. See Reducing Flame.

Carbon Monoxide. A colorless, odorless toxic gas, CO, a product of incomplete burning of carbon.

Carbon Steel. Low carbon or mild steel.

Carbon Tetracholride. Colorless nonflammable and very toxic liquid used as a solvent; it should never be allowed to touch skin and fumes must not be inhaled.

Carborundum. Trademark. A compound of carbon and silicon used as an abrasive.

Carcinogen. Any substance that produces cancer.

Card Access Control. An entry device that is operated by a small magnetized plastic card, similar to a credit card.

Carillon. A set of bells sounded either from a keyboard or manually.

Carnuba Wax. A hard wax obtained from species of palm grown mostly in Brazil.

Carpenter. A craft worker skilled in woodwork, especially of the structural kind.

Carpenter Pencil. A sturdy pencil with a thick rectangular lead for marking lumber.

Carpenter's Level. A hand tool consisting of a wood or metal bar with spirit levels attached, used for establishing plumbness and levelness of construction members.

Carpenter's Saw . See Handsaw.

Carpentry, Finish. The finish woodwork installation such as base, casing, doors, stairs, paneling; all woodwork installed after plastering or drywall.

Carpentry, Rough. The preliminary framing, boxing, and sheeting of a wood frame building.

Carpet. General designation of fabric constructions which serve as soft floor coverings, especially those which cover the entire floor and are fastened to it, as opposed to rugs; see also Woven, Tufted, Knitted, Punched, and Flocked carpets.

Carpet Base. 1. Vinyl or rubber base attached to the wall with adhesive and installed as a finish for carpeting. 2. A base made of carpeting serving the same purpose.

Carpet Lining. See Carpet Padding.

Carpet Padding. A jute, felt ,hair, foam, or plastic rubber underlayment installed under carpeting to increase underfoot comfort, to absorb pile-crushing forces and to reduce impact sound transmission; also called Cushioning, Lining, or Underlay.

Carpet Plate. A thin piece of ornamental metal that the rough edges of carpet is connected to at doorways or at the junction of carpet to another flooring material to form a clean ending point.

Carpet strip. See Base Shoe.

Carpet Tack. A small tack used for attaching carpeting to wood floors.

Carpet Tile. Carpet that comes in sheets or small squares and is installed with the use of adhesives.

Carport. An open-sided shelter with a roof for a car, usually beside a house.

Carrel. A small cubicle for an individual reader in a library.

Carrene. Refrigerant in Group One (R-11); chemical combination of carbon, chlorine and fluorine.

Carriage. The framing support of the steps and risers of a flight of stairs; also called Rough Horse and Stair Horse.

Carriage Bolt. A bolt with a round smooth head that is threaded; a bolt with a square neck directly under the head to prevent its rotation.

Carrier Channel. The main supporting metal members used in the construction of suspended ceilings.

Carrier, Fixture. A mounting frame built into a wall to support a plumbing fixture.

Cart, Laundry. A push or pull institutional vehicle with wheels, for the transport of clean or soiled laundry.

Cartouche. An ornamental frame.

Cartridge. 1. Any of various small containers, holding a supply of material for a larger device into which it is inserted. 2. Disposable element containing filtering media and used in some pool filters.

Cartridge, Extinguisher. The cylindrical container of a fire prevention apparatus which contains the chemicals used in the suppression of fire.

Cartridge Filter. A swimming pool filter which operates through a disposable cartridge; these are of two general types. the surface or area type where the suspended matter is removed at the surface and the depth type in which the interstices vary from large to small in depth.

Carved Carpet. See Sculptured Carpet

Carved Door. A door that has been finished with either factory or hand cutting onto its surface to improve its appearance.

Cascade Sequence. In welding, a combined longitudinal and buildup sequence wherein weld beads are deposited in overlapping layers; in manual shielded metal arc welding a backstep sequence is normally used. See Block Sequence and Buildup Sequence.

Cascade System. Arrangement in which two or more refrigerating systems are used in series; uses the evaporator of one machine to cool the condenser of another machine; produces ultra-low temperatures.

Casein. Protein obtained from milk, soluble in alkaline water solution; casein is used extensively in the manufacture of water paints.

Case Joint. A type of joint used in cabinetmaking in which the two pieces are butted together at an angle and fastened by dowels.

Casement. A window in which the sash opens with hinges and pivots on an axis along the vertical line of the frame; casement window.

Case Mould. Plaster shell used to hold various parts of a plaster mould in correct position; also used with gelatin and wax moulds to prevent distortions during pouring operation.

Case, Refrigerated. A storage case kept cold by a mechanical device. For use to store perishable items.

Casein Glue. An adhesive substance composed of casein (the curd of milk), lime, and sodium salt; it comes as dry powder to which water is added.

Casein Paints. Paint in which casein solution has replaced the binder.

Casework. Assembled cabinetry or millwork.

Cash Accounting. A method of keeping accounting records in which income is recorded when actually received and expenses are recorded when cash is paid out; also called Cash Basis.

Cash Basis. See Cash Accounting.

Cash Flow. The actual cash income after all cash outlays and reserves have been deducted from the gross income.

Casing. 1. The wood finish pieces surrounding the frame of a window or door, or the finished lumber around a post or beam. 2. A cylindrical steel tube used to line a drilled or driven hole such as a well or caisson.

Casing Bead. Metal or wood molding used to separate different materials, used as an edge or used around openings to provide a stop.

Casing, Ranch. An architectural style of exposed millwork enclosure of cased beams, posts, pipes, and the exposed molding or lining around doors and windows.

Casing Trim. Metal or wood material that is attached around windows and doors to act as the decorative finish.

Casino. 1. A building or room used for gambling or other amusements. 2. A small summerhouse

CASSB. Cedar Shake & Shingle Bureau.

Cassiterite. An inorganic mineral of the tetragonal form used as a source of tin and tin oxide; SnO_2.

Cast. Inclination of one color to look like another; for example, sulphur is yellow with a greenish cast.

Castellated Beam. A steel, wide-flange section whose web has been cut along a zigzag path and reassembled by welding in such a way as to create a deeper section.

Casting. Something cast in a mold, like cast iron or cast aluminum.

Casting Bed. A form in which precast concrete units are constructed.

Casting Mold. Use of gelatin, wax, or plaster molds to make plaster ornamentation

Casting Plaster. A fast-setting gypsum plaster that is used to anchor marble to walls; see Gypsum Molding Plaster.

Casting, Solid. Forming castings by introducing a body slip into a porous mold which usually consists of two major sections, one section forming the contour of the inside of the ware and allowing a solid cast to form between the two mold faces.

Cast-In-Place Concrete. Concrete that is poured in its intended location at a site.

Cast-In-Place. Mortar or concrete which is deposited in the place where it is required to harden as part of the structure, as opposed to precast concrete.

Cast Iron Wheel Guard. Lineal component placed at intersection of wall and horizontal surface to restrain wheels of vehicles from coming close to wall surface, protecting it from vehicular damage.

Cast Iron. Iron with a high carbon content, which cannot, because of the percentage of carbon, be classified as steel.

Castor Oil. Nondrying oil obtained from the castor bean; may be converted to a drying oil by chemical treatment.

Casts. Finished plaster products from a mold, sometimes referred to as staff; used generally as enrichments and stuck in place.

Cast Stone. Concrete cast in molds for ornamental use in construction.

Catalyst. A substance that initiates a chemical reaction and enables it to proceed under different conditions (as at a lower temperature) than otherwise possible.

Catch. A piece of hardware for fastening a door, window, or cabinet door.

Catch Basin. A receptacle for catching water runoff from a designated area; usually a shallow concrete box with a grating and a discharge pipe leading to a plumbing or stormwater system.

Catenary. The curve assumed by a completely flexible string or cable loaded only by its own weight.

Catface. Blemish or rough depression in the finish coat of plaster caused by variations in base coat thickness.

Cathedral. A church that is the official seat of a diocesan bishop.

Cathode. 1. The negative electrode in an electrolytic cell. 2. The positive terminal of a primary cell such as a battery.

Cathode-Ray Tube. A high-vacuum tube in which cathode rays produce a luminous image on a fluorescent screen.

Catwalk. A narrow walkway, such as used in an attic for access.

Caulk. 1. To seal and waterproof cracks and joints, especially around window and exterior door frames. 2. To seal small openings in wall or ceiling systems to prevent leakage of sound or to effect a finished appearance and seal between dissimilar materials.

Caulking. 1. A composition of vehicle and pigment, used at ambient temperatures for filling joints; remains plastic for an extended time after application. 2. A method of making a bell and spigot pipe joint watertight by packing it with oakum and lead or other materials.

Caulking Compound. A soft, plastic material used for sealing joints in buildings and other structures where normal structural movement may occur; retains its plasticity for an extended period after application; available in forms suitable for application by gun and knife and in extruded preformed shapes.

Caulking Gun. A device, usually hand-powered, which dispenses liquid caulking into joints and seams.

Caulk, Masonry. A resilient material applied where masonry work butts against other materials to seal cracks or openings.

Caustic Lime. Lime.

Caveat. A warning or proviso.

Caveat Emptor. Let the buyer beware. A warning that the buyer purchases at its own risk

Cavetto. A quarter hollow molding, the converse of a quarter round

Cavitation. Localized gaseous condition or partial vacuum that is found within a liquid stream; caused by mechanical force, as in a pump impeller, or in fluids at high velocities.

Cavity. A hollow or void space within a mass.

Cavity Wall. A masonry wall that includes a continuous airspace between its outermost wythe and the remainder of the wall.

Cavity Wall Anchor. A metal device mounted on a masonry wall used to secure other attachments or masonry to an existing or back-up wall.

CBD. Commerce Business Daily; a daily newspaper that carries listings of governmental contracts available for bidding.

C/B Ratio. The ratio of the weight of water absorbed by a masonry unit during immersion in cold water for 24 hours to weight absorbed during immersion in boiling water for 5 hours; an indication of the probable resistance of brick to freezing and thawing; also called saturation coefficient.

CC&Rs. Covenants, conditions, and restrictions.

C Channel. A C shaped steel or aluminum section shaped like a rectangular box with one side removed.

C-Clamp. A clamp in the shape of a "C" with jaw capacities usually ranging from 1 to 8 inches used for the securing of wood or metal pieces in a fixed position and for temporary assemblies.

CCMCA. California Conference of Mason Contractor Associations, Inc.

CCTV Cable. Cable that is used for the transmission of closed-circuit television.

CCTV. Closed Circuit Television.

CDA. Copper Development Association.

CDX Plywood. Plywood used in exterior applications that is graded C and D, for sheathing.

CDX. A grading system mark for plywood which means. grade C and D, exterior glue.

CE. Civil Engineer.

Cedar. An aromatic durable softwood, genus Cedars, of the pine family.

Cedar Closet. A closet that is lined with thin pieces of cedar wood; used for its fragrance and its ability to repel insects.

Cedar Deck. A flat-floored roofless area adjoining a structure constructed of cedar wood; a platform serving as a structural element constructed of cedar wood; planks for flooring, from cedar, usually 2" nominal thickness.

Cedar Shake. A shingle made by splitting a block of cedar along its grain and thereby creating a shingle which may be used for roofing or siding; also called Handsplit Shingles.

Cedar Shingle. A thin piece of cedar wood with one end thicker than the other for laying in overlapping rows as a covering for a roof or the sides of a building or structure.

Cedar Siding. Boards milled from cedar wood, used for the finish covering on the exterior walls of a building or structure; used for its appearance and resistance to moisture and aging.

Ceiling. 1. The overhead inside lining of a room; classified by structural type, contact, furred, or suspended.

Ceiling Access Door. A hinged door or loose fitting panel that allows for admittance to an attic.

Ceiling Blocking. Wood pieces installed between ceiling joists and rafters to provide nailing surfaces for finishing ceiling materials.

Ceiling Diffuser. A mechanical device through which warm or cold air is blown into an enclosure, for the purpose of to distributing conditioned air.

Ceiling Framing. Wood or metal pieces which form the rough framing of ceilings.

Ceiling Furring. Wood or metal strips applied to a ceiling or rafter to make the ceiling or rafter level, provide a nailing surface, or create an air space.

Ceiling Grille. A grating, screen, or louvered panel that allows air into a ventilating duct.

Ceiling Heater. An electric heater installed in a ceiling, often in a bathroom.

Ceiling Insulation. Loose, blown-in material or fiberglass rolls that are in installed at the ceiling plane.

Ceiling Joist. The horizontal members in a building or structure to which the ceiling material is fastened.

Ceiling Lath. Sheets of expanded metal, gypsum or in older structures, wood lath, which are attached to a ceiling to provide a plaster base.

Ceiling Molding. Molding that is used to form a projection at the top of a wall.

Ceiling Mortar. Extra-rich wall mortar.

Ceiling Painting. The actual physical process of applying paint, either by brush, roller, or spray gun to the ceiling section of a structure.

Ceiling Plenum. Space below the flooring and above the suspended ceiling that accommodates the mechanical and electrical equipment and that is used as part of the air distribution system.

Ceiling Price. The maximum price that an informed buyer would pay to purchase or lease property.

Ceiling Removal. The demolition and removal of ceiling materials in order to replace or remodel.

Ceiling Sound Transmission Class. A measure of reduction in sound transmission via plenum path between two rooms.

Cell. 1. Any void space. 2. A single room in a prison or jail. 3.The anatomical units of plant tissue, including wood fibers, vessel members, and other elements of diverse structure and function. 4. One of the hollow openings in building tile or cement blocks. 5. In electrical raceways, a single, enclosed tubular space in a cellular metal floor member, the axis of the cell being parallel to the axis of the metal floor member.

Cellar. Basement

Cellular Concrete. A lightweight product consisting of portland cement, cement-pozzolan, cement-sand, lime-pozzolan or lime-sand pastes, or pastes containing blends of these ingredients and having a homogenous void or cell structure, attained with gas- forming chemicals or foaming agents; for cellular concretes containing binder ingredients other than or in addition to portland cement, autoclave curing is usually employed; also called Foam Concrete or Gas Concrete.

Cellular Decking. Metal floor or roof deck panels made of steel sheets corrugated and welded together in such a way that hollow longitudinal cells are created within the panels.Deck which during construction supports wet concrete and construction loads, but after concrete cures does not perform structural function in completed construction; deck also is fabricated of two sheets to form linear voids.

Cellular Raceway. The hollow spaces of cellular metal floors, together with suitable fittings, that are used as enclosures for electrical and telephone conductors.

Cellulose. 1. The carbohydrate that is the principal constituent of wood and forms the framework of the wood cells. 2. An organic substance obtained from the cotton plant and used as raw material in the manufacture of paints and other materials.

Cellulose Acetate. A binder made by chemical reaction of acetic acid on cellulose (cotton linters).

Cellulose Nitrate. A binder made by chemical reaction of nitric acid on cellulose (cotton linters); also called Nitrocellulose or Pyroxylin.

Celsius. International thermometric scale where 0.01 degrees represents the triple point of water and 100 degrees the boiling point; similar to Centigrade.

Celsius Temperature Scale. The temperature scale used in metric system in which the freezing point of water is 0° and the boiling point is 100°; see Celsius.

Cement Asbestos. A material composed of portland cement, fine aggregate, and asbestos fibers; it is formed into flat and corrugated building boards used for roofing and siding, pipes and fittings, and water tanks.

Cement Base Paint. A paint composed of portland cement, lime, pigment, and other modifying ingredients; sold as dry powder to be mixed with water for application.

Cement Body Tiles. Tiles with the body made from a mixture of sand and portland cement; the surface may be finished with portland cement, spheroids of marble, or other materials.

Cement Color. Colored powdered or liquid pigments added to a mix to integrally color concrete.

Cement Content. The quantity of cement contained in a unit volume of concrete or mortar, ordinarily expressed as pounds, barrels, or bags per cubic yard.

Cement Factor. The number of bags or cubic feet of cement per cubic yard of concrete; see Cement Content.

Cement Fiber Board. A prefabricated concrete building sheet that is compressed and bonded.

Cement Fiber. A threadlike structure added to cement to stiffen and strengthen it.

Cement Gel. The colloidal, glue like, material that makes up the major portion of the porous mass of which hydrated cement paste is composed.

Cement Grout. 1. A cementitious mixture of portland cement, sand or other ingredients, and water which produces a uniform paste used to fill joints and cavities between masonry units. 2. A thin mortar used for pointing-up and finishing joints between tile units.

Cementing. In roofing, a solidly mopped application of hot asphalt, cold liquid asphalt compound, hot coal-tar pitch, or other cementing material.

Cementitious. Having cementing properties; usually used with reference to inorganic substances, such as portland cement and lime.

Cementitious Material. A component material of plaster, mortar, or concrete which when mixed with water provides plasticity necessary for placement; upon subsequent setting or hardening it serves to bind aggregate particles together into a rigid heterogeneous mass.

Cementitious Topping. A compound that is capable of setting like concrete when applied on a concrete base to form a floor surface.

Cement, Keene's. See Keene's Cement.

Cement, Masonry. A hydraulic cement for use in mortars for masonry construction, containing one or more of the following materials. portland cement, portland blast-furnace, slag cement, portland-pozzolan cement, natural cement, slag cement or hydraulic lime; and in addition usually containing one or more materials such as hydrated lime, limestone, chalk, calcereous shell, talc, slag, or clay, as prepared for this purpose.

Cement Mortar. A mixture of cement, lime, sand, or other aggregates, and water, used for plastering over masonry or to lay block, brick or tile.

Cement Paste. The mixture of portland cement, water, pozzolans and other admixtures, if any, and air which surround the aggregates in concrete; also called the matrix.

Cement Plaster. 1. Plaster having portland cement as its binder; used on exterior surfaces or in damp areas. 2. Gypsum plaster made to be used with the addition of sand for basecoat plaster; also called Neat or Hardwall plaster.

Cenotaph. A tomb or monument erected in honor of a person or group of persons whose remains are elsewhere.

Center. 1. The middle point of a line, circle, or sphere, equidistant from the ends or from any point on the circumference or surface; a pivot or axis of rotation. 2. A place or group of buildings forming a central point in a district or city. 3. A temporary structure to support the arch while it is being built; see Centering.

Center-Hung Sash. A sash hung on its centers so that it swings on a horizontal axis.

Centering. Temporary formwork for an arch, dome, vault, or other overhead surface.

Centering Shims. Small blocks of synthetic rubber or plastic used to hold a sheet of glass in the center of its frame.

Centerline. A real or imaginary line that is equidistant from the sides of some object; it is usually represented on drawings as a line of alternate dots and dashes.

Center Matched. See Tongue and Groove.

Center of Gravity. The point at which the weight of a body may be considered to act; center of mass.

Center Pole. Column in center of spiral stair which supports stair treads.

Center Punch. A hand punch consisting of a short steel bar with a hardened conical point at one end used for marking the centers of holes to be drilled.

Center to Center (C to C). The dimension from the centerline of one member to the centerline of the next member.

Centesimal Measure. 1. Division into hundredths. 2. Division of the circle into 400 grads.

Centigrade. Thermometric scale where 0 degrees represents the freezing point of water and 100 degrees the boiling point. Similar to Celsius.

Centimeter. A metric unit that equals one-hundredth of a meter or 10 millimeters and is equivalent to 2.54 inches.

Centipose. A metric unit of viscosity.

Central Inverter. A device for converting direct current into alternating current by mechanical or electronic means.

Central System. A system of conditioning air supplied to various areas or space, serviced by the same source of heat or cooling; all equipment in central systems is indoors except air-cooled condensers, evaporative condensers, and cooling towers.

Centrifugal. Away from the center; opposite from centripetal.

Centrifugal Compressor. Pump which compresses gaseous refrigerants by centrifugal force.

Centrifugal Force. An apparent force that acts outwards on a body moving about a center.

Centrifugal Pump. A pump which draws water into the center of a high speed impeller and forces the fluid outward with velocity and pressure.

Centrifuge, Laboratory. A laboratory apparatus using centrifugal force for separating substances of different densities, for removing moisture, or for simulating gravitational effects.

Centripetal. Towards the center; opposite from centrifugal.

Centripetal Force. A force that keeps a body moving about a center from flying outwards.

Centroid. Center of mass.

Ceramic. Made of clay and permanently hardened by heat.

Ceramic Adhesive. Used for bonding tile to a surface; rubber solvents; rubber- and resin-based emulsions used as adhesives.

Ceramic Color Gaze. An opaque colored glaze of satin or gloss finish obtained by spraying the clay body with a compound of metallic oxides, chemicals and clays; it is burned at high temperatures, fusing glaze to body, making them inseparable.

Ceramic Insulator. A device made of ceramic non-conductive material which is used in electrical installations.

Ceramic Mosaic Tile. An unglazed tile formed by either the dust-pressed or plastic method, usually 1/4 to 3/8 in. (6.4 to 9.5 mm) thick, and having a facial area of less than 6 in, usually mounted on sheets approximately 2 by 1 ft. (0.3 by 0.6 m) to facilitate setting; ceramic mosaic tile may be of either porcelain or natural clay composition and may be either plain or with an abrasive mixture throughout.

Ceramic Process. The production of articles or coatings from essentially inorganic, nonmetallic materials, the article or coating being made permanent and suitable for utilitarian and decorative purposes by the action of heat at temperatures sufficient to cause sintering, solid-state reactions, bonding, or conversion partially or wholly to the glassy state.

Ceramics. A general term applied to the art or technique of producing articles by a ceramic process, or to the articles so produced.

Ceramic Tile. A thin surfacing unit made from clay and/or a mixture of clay and other ceramic materials; the tile has either a glazed or unglazed face; it is fired above a red heat in the course of manufacture to a temperature sufficiently high to produce specific physical properties and characteristics.

Ceramic Veneer. A type of architectural terra cotta, characterized by larger face dimensions and thinner sections ranging from 1-1/8 in. to 2-1/2 in. in thickness.

Ceresin. A hydrocarbon wax which possesses considerable flexibility.

Certificate for Payment. A written document forwarded to the general contractor by the architect, engineer, or owner approving payment for work completed.

Certificate of Insurance. A certificate provided by the general contractor verifying that he has obtained the required insurance for the project; the certificate is issued by the insurance company or its agent and confirms the existence of the insurance, the coverage, and its expiration date.

Certificate of Substantial Completion. A written document forwarded to the general contractor by the architect, engineer, or owner indicating that the project is substantially complete; this document initiates the time period for the final payment to the contractor.

Certified Check. A depositor's check guaranteed for payment by the bank.

Cessation of Work. The ending of work on a construction project without completion.

Cesspool. A subterranean container for temporary storage of septic tank effluent while it soaks into the adjoining soil.

CFC. Chlorofluorocarbon.

CFM. Cubic feet per minute.

CFS. Cubic Feet per Second.

CGL Insurance. Comprehensive General Liability Insurance.

Chain. A flexible series of connected metal links, to support a load.

Chain Binders. In carpet making, yarns running warpwise (lengthwise) in the back of the carpet, binding all construction yarns together; the chain binder runs alternately over and under the weft binding and filling yarns, thereby pulling the pile yarn down and the stuffer yarns up for a tightly woven construction.

Chain Hoist, Door. A chain in a grooved pulley or sheave with a chain hook used to hoist a large door.

Chain Link Fence. A fence made of a wire mesh fabric.

Chain Trencher. A self-propelled machine with blades attached to a continuous chain, used to excavate trenches.

Chain Warp. In carpet making, zigzag warp yarn that works over and under the shot yarns of the carpet, binding the backing yarns together; see Warp.

Chair. 1. A separate seat for one person, usually with four legs and a back. 2. A device used to support reinforcing bars while concrete is being poured.

Chair, Hydrotherapy. A device for the immersion of a medical patient in water.

Chair, Lifeguard. A raised chair, equipped with a ladder, that affords a lifeguard an elevated view of a large area.

Chair Rail. A wood molding separating the dado or wainscot from the upper wall; usually at a convenient height to prevent chair backs from abrading the wall.

Chair, Reinforcing. Metal supports made of fabricated wire, made to hold reinforcing steel in place until concrete is poured.

Chalet. 1. A Swiss alpine dwelling with exposed structural elements and wide roof overhangs on front and sides. 2. A small suburban house or bungalow, in the chalet style, particularly with a broad roof overhang.

Chalk. 1.A form of natural calcium carbonate; see whiting. 2. A lump of soft limestone used by carpenters for impregnating a snap line. 3. A crayon for marking materials on a construction site or in a workshop.

Chalkboard. Panel for writing on with chalk or liquid chalk.

Chalking. 1. The decomposition of a paint film into a loose powder on the surface; mild chalking, accompanied by satisfactory color retention in tinted paint, is considered a desirable characteristic; heavy chalking which washes off to leave an unprotected surface is highly undesirable; before recoating a heavily chalked surface, all of the chalk should be removed by vigorous brushing. 2. The dusty powdering on an asphalt roof surface that is subject to ultraviolet degradation.

Chalk Line. A straight working line made by snapping a chalked cord between two points; also called a Snap Line; see Chalk Reel.

Chalk Rail. A trough mounted under a chalkboard to store chalk and erasers and to catch the chalk dust.

Chalk Reel. A carpenter's implement consisting of a string reel and chalk in a container, used as a method of chalking a snap line and storing the string.

Chamber. A room or space; a bedroom.

Chamfer. A beveled surface cut on the corner of a piece of wood.

Chamfering Strip. Piece of stock placed in an inside corner of a form to produce a beveled edge.

Chamfering Strip. See Chamfer Strip.

Chamfering. The preparation of a contour other than for a square groove weld on the edge of a member for welding.

Chamfer Strip. An insert that is triangular or curved, placed in an inside corner to produce a rounded or flat beveled edge at the right angle corner of a construction member; also called Chamfering Strip.

Chandelier. An ornamental branched hanging fixture for several candles or electric light bulbs.

Change Of State. Condition in which a substance changes from a solid to a liquid or a liquid to a gas caused by the addition of heat, or the reverse, in which a substance changes from a gas to a liquid, or a liquid to a solid, caused by the removal of heat.

Change Order. An order to change the work to be performed under a construction contract, usually given by an owner to a prime contractor or a by prime contractor to a subcontractor; a revision in the contract documents after the execution of the owner-contractor contract.

Changes. Key changes.

Change Trailer. A temporary vehicle that is used by personnel on a jobsite.

Channel. A U-shaped rolled steel or extruded aluminum section shaped like a rectangular box with one side removed.

Channel Block. 1. A concrete masonry unit with a solid bottom and sides but no ends or webs, for use in a lintel. 2. A hollow unit with web portions depressed less than 1-1/4 inches to form a continuous channel for reinforcing steel and grout.

Channel Closure Strip. A U-shaped resilient strip used to close openings in metal panels and flashing.

Channel Door Frame. A U-shaped or L-shaped channel used as a door frame.

Channel Furring. A formed sheet metal furring strip.

Channel, Manhole. The bottom of a sanitary or stormwater manhole which creates a channel between the incoming and outgoing pipes.

Channels. 1. In asphalt paving, ruts or grooves that may develop in the wheel tracks of a pavement; may result from consolidation or lateral movement under traffic in one or more of the underlying courses, or by displacement in the asphalt surface layer itself; they may develop under traffic in new asphalt pavements that had too little compaction during construction or from plastic movement in a mix that does not have enough stability to support the traffic. 2. Hot or cold-rolled steel, of various sizes, used for furring, studs and in suspended ceilings.

Channels, Carrying. The heaviest integral supporting member in a suspended ceiling; carrying channels, or main runners, are supported by hangers attached to the building structure, and in turn, support various grid systems and furring channels or rods to which lath is fastened.

Channels, Furring. The smaller horizontal member of a suspended ceiling, applied at right angles to the underside of carrying channels and to which lath is attached; the smaller horizontal member in a furred ceiling; in general; the separate members used to space lath from any surface member over which it is applied.

Channel Slab. A manmade watercourse of molded, layered, plain or reinforced concrete.

Channel Strap. A U-shaped or L-shaped iron plate used to connect two or more timbers.

Channel Valve. A valve which controls the flow of water from a natural or artificial water course.

Chapel . 1. A place for private worship in a large church or attached to a house or institution. 2. A room for services in a funeral home.

Character. A requirement by a loan officer to evaluate experience with similar jobs and locations, business reputation with lenders, suppliers, and subcontractors and reasonableness of bid.

Charcoal. An amorphous form of carbon consisting of a porous black residue from partially burnt wood, bones, or other substances; made by charring in a kiln from which air is excluded.

Charge. Amount of refrigerant placed in a refrigerating unit.

Charging Board. Specially designed panel or cabinet fitted with gauges, valves, and refrigerant cylinders used for charging refrigerant and oil into refrigerating mechanisms.

Charles' Law. For a constant pressure, the volume varies directly as the absolute temperature, and for constant volume, pressure varies directly as absolute temperature.

Chase. A groove or indentation cut into masonry to accommodate electric or plumbing lines.

Chattel. Moveable personal property.

Check. A lengthwise separation of the wood that usually extends across the rings of annual growth and commonly results from stresses set up in wood during seasoning.

Check Cracks. See Craze Cracks.

Checking. 1. The pattern of irregular surface cracks on the top pour of an asphalt roof, a preliminary stage of alligatoring. 2. Cracks or fissures that appear with age in many exterior paint coatings; superficial at first, they may in time penetrate the coating entirely; these cracks may assume many patterns, but the usual ones resemble the print of a bird's foot or small squares.

Checkrail. A meeting rail sufficiently thicker than a window sash to fill the opening between the top and bottom sash made by the parting stop in the frame of double-hung windows; checkrails are usually beveled.

Checkroom Shelf. A horizontal mounted surface upon which objects can be stored and kept track of in a checkroom.

Check Valve. A device which allows fluid or air to pass through in only one direction; a valve which prevents the back-flow of water or other liquid by automatically closing.

Chemical Bond. 1. The bond produced by cohesion between separate laminates of similar crystalline materials; based on formation and subsequent interlocking of crystals. 2. The adherence of one plaster coat to another or to the base which implies formation of interlocking crystals or fusion between the coats or to the base.

Chemical Brown Stain. A chemical discoloration of wood which sometimes occurs during the air drying or kiln drying of several species apparently caused by the concentration and modification of extractives.

Chemical Extinguisher. A wheeled device or hand-held cylinder that contains chemicals to extinguish certain types of fires.

Chemical Feed. Injection of chemicals into pool water circulation for pollution control.

Chemically Active. 1. Of paint pigments, those which react with oil of vehicle to form soaps which influence toughness of film, and increase durability. 2. Pigments such as red lead which react with acids formed at metal surface to prevent rust.

Chemically Pure (CP). Of the highest grade but not necessarily 100 percent pure.

Chemical Piping. Piping which conveys concentrated chemical solutions from a feeding apparatus to the circulation piping.

Chemical Porcelain. Vitreous ceramic whitewares used for containing, transporting, or reacting of chemicals.

Chemical Refrigeration System. System of cooling using a disposable refrigerant; also called an expendable refrigerant system.

Chemical Resistance. The effect of specific chemicals on the properties of various materials with respect to concentration, temperature and time of exposure.

Chemical Toilet. A self-contained portable toilet, not attached to a sewer line, for temporary use on a construction site.

Chenille. A pile fabric woven by the insertion of a prepared weft row of surface yarn tufts in a fur or caterpillar form through very fine but strong cotton catcher warp yarns, and over a heavy woolen backing yarn.

Cherry Veneer. A thin layer of cherry wood used as a finished surface material.

Chestnut Veneer. A thin layer of chestnut wood used as a finished surface material.

Chevron Bracing. That form of bracing wherein a pair of braces located either above or below a beam terminates at a single point within the clear beam span.

Chicken Ladder. A lightweight ladder that can be hung over the ridge for work on a steep roof.

Chicken Wire. Thin, galvanized, hexagonal, woven wire mesh mounted on an exterior wall as a base for stucco plaster.

Chilled Water System. A cooling system in which the entire refrigeration cycle occurs within a single piece of equipment; water is used to bring the heat from the space to the evaporator section of the chiller, and water is also used to carry the heat from the condenser to the outside.

Chiller. A piece of equipment that produces chilled water for circulation through a building and contains a compressor, condenser and evaporator tank.

Chiller, Absorption. A water cooling system similar to a vapor compression chiller with the exception that it does not use a compressor, but uses thermal energy - low pressure steam, hot water, or other hot liquids to produce the cooling effect.

Chill Factor. See Wind-chill.

Chime. 1. An apparatus for chiming a bell or a set of bells. 2. An electrical device used as a door bell, actuated by a push button.

Chimney. A vertical, noncombustible structure with one or more flues to carry smoke and other gases of combustion into the atmosphere.

Chimney Breast. The projecting portion of a chimney, especially when projecting from the exterior wall of a building or into a room.

Chimney Brick. Brick used for the construction of chimneys because of its ability to withstand high temperatures without cracking.

Chimney Connector. Pipe connecting a heating appliance, such as a furnace, with the vertical flue.

Chimney Effect. Tendency of air or gas to rise when heated.

Chimney Flue. A channel or shaft in a chimney for conveying smoke and exhaust gases to the exterior atmosphere.

Chimney Lining. Fire clay or terra cotta material, or refractory cement, made to be built inside of a chimney.

Chimney, Masonry. A vertical noncombustible structure with a flue or flues to remove smoke and other gases, constructed of shaped or molded masonry units.

Chimney Pot. An earthenware or metal pipe at the top of a chimney, narrowing the aperture and increasing the updraft.

Chimney Throat. The narrowest part of a fireplace chimney, adjacent to the smoke shelf; the location of the damper.

China. A glazed or unglazed vitreous ceramic whiteware used for nontechnical purposes; designation of such products as dinnerware, sanitary ware, and art ware when they are vitreous.

China Clay. Kaolin.

China Process. The method of producing glazed ware by which the ceramic body is fired to maturity, following which the glaze is applied and matured by firing at a lower temperature.

China Wood Oil. Tung oil.

Chinese Blue. A form of iron blue.

Chinese Red. Chrome orange, deep.

Chipboard. A paperboard used for many purposes that may or may not have specifications for strength, color, or other characteristics; normally made from paper stock with a relatively low density in the thickness of 0.006 inch and up.

Chip Cracks. See Eggshelling.

Chipped. In tile work, caused by rough handling and confined to the corners and edges of the tile; the scaling or breaking off at the edges of fragments from the surface of a tile.

Chipping. 1. Removing welding defects and surface slag by use of a chipping chisel. 2. Cleaning steel using special hammers. 3. Type of paint failure.

Chipping Hammer. A mason's hand tool, often capped with tungsten carbide for durability; used to chip excess material from the backs and edges of block, brick, stone, or tile.

Chisel, Cold. See Cold Chisel.

Chisel Edge. A slanted factory edge on gypsumboard.

Chisel, Wood. See Wood Chisel.

Chloramine. See Ammonia.

Chlordane. A chlorinated substance used as a pesticide.

Chlorinated Isocyanurate. Chlorine and cyanuric acid compound used to maintain chlorine level in pool water and prevent chlorine from dissipating in sunlight. See Conditioned Water.

Chlorinated Polyvinyl Chloride. A type of plastic used to make pipe to carry hot water and chemicals.

Chlorinated Rubber. A particular film former used as a binder, made by chlorinating natural rubber.

Chlorine. A poisonous greenish-yellow gaseous element of the halogen group, used for purifying water, bleaching, and the manufacture of many organic chemicals.

Chlorine Demand. The amount of chlorine necessary to oxidize all organic material present in pool water at a given moment or over a period of time.

Chlorine Residual. The amount of chlorine remaining in pool water after the chlorine demand has been satisfied at a given moment; this chlorine is available to oxidize other bacteria in water.

Chlorofluorocarbon. A compound that contains carbon, chlorine, and fluorine; known commonly as CFCs, they were widely used in refrigerators and aerosol sprays, but it is now known that they have harmful effects on the earth's atmosphere.

Chock. Heavy timber or wooden block, fitted under tires or wheels to prevent movement.

Choir. The part of a cathedral or large church between the altar and the nave.

Choir Loft. A gallery or balcony in a church to be occupied by a choir.

Choke Tube. Throttling device used to maintain correct pressure difference between high-side and low-side in refrigerating mechanism. Capillary tubes are sometimes called choke tubes.

Chopper, Food. A device that chops food and blends it into smaller pieces.

Chord. 1. One of the main members of a truss, braced by web members of the truss. 2. Perimeter member of a building or structure which resists lateral forces.

Chroma. Saturation, purity, or intensity of color.

Chrome. Chromium.

Chrome Green. Mixture of chrome yellow and Prussian blue, one of industry's most important green pigments.

Chrome Orange. An orange pigment composed principally of basic lead chromate.

Chrome Yellow. Important inorganic yellow pigment made by mixing solutions of lead acetate and potassium bichromate; highly corrosion-inhibiting.

Chromium. A blue-white metallic element, used as a shiny decorative electroplated coating.

Chromium Oxide Green. Green pigment which is extremely permanent in color and has good resistance to both alkali and heat.

Chromium Oxide. See Chromium Oxide Green.

Chromometer. An instrument used to indicate the color of light liquids and oils; also called a Colorimeter.

Chronometer. A timepiece; a clock.

Chuck. An attachment for holding a tool in a machine, as a bit on a drill.

Chute. An inclined plane, sloping channel, or passage down or through which materials may pass.

Chute, Mail. See Mail Chute.

Chute, Trash. See Trash Chute.

CI. Cast iron.

Circle Cutter. An adjustable scribe tool for cutting circular patterns or openings for lighting fixtures and other devices in gypsum board.

Circuit. The path of an electric current from the source.

Circuit Breaker. An overcurrent protection device.

Circuit, Parallel. See Parallel Circuit.

Circuit, Pilot. See Pilot Circuit.

Circuit Protector. A device that will open an electrical circuit in the event of an overload, thus protecting operating equipment and other components from damage.

Circuit, Series. See Series Circuit.

Circuit Vent. This is a group plumbing vent that extends from the front of the last fixture of the horizontal group to the vent stack; this type of vent may be used when the circuit carries the drainage of from two to eight urinals, water closets, stall showers, or sinks.

Circular Mil. A unit of length used in wire sizes.

Circular Saw. A power saw with a circular cutting blade, either in the form of a portable hand tool or a stationary table saw.

Circulating Fireplace. A fireplace that has cold air intakes and hot air outlets into and out of a heat exchanger that is built into the firebox to enhance the effectiveness of a fireplace as a heating system.

Circulation Path. An exterior or interior way of passage from one place to another for pedestrians, including, but not limited to, aisles, walks, hallways, courtyards, stairways, and stair landings.

Circulation Piping System. The piping between a pool, spa or hot tub structure and the mechanical equipment; usually includes suction piping, face piping and return piping.

Circulation Pump. A pump that moves fluids in a piping system such as in a domestic hot water system or in a hot water heating system.

Circulation System. Entire flow arrangement of fittings, pipework, and equipment.

Circumference. The enclosing boundary of a circle or other figure enclosed by a curve.

CISPI. Cast Iron Soil Pipe Institute.

Cistern. A tank for storing water.

Citronella Oil. An oil with a peculiar odor, obtained from a species of grass grown in Asia.

City Hall. A building housing city government administrative offices.

Civil Engineer. An engineer who designs public works such as roads, harbors, piping, earthwork, and waterworks.

Cladding. 1. Metal exterior building surfacing panels. 2. In welding, see Surfacing.

Cladding Panel. A panel applied to a structure to provide durability, weathering, corrosion and impact resistance.

Clamp. A mechanical device used to hold two or more pieces of material together.

Clamp, Beam. A device which holds a horizontal structural member to a vertical member.

Clamp, Ground Rod. The device that attaches the main ground wire to a cold water pipe in an electrical system.

Clapboard. A type of wood siding consisting of narrow boards thicker on one side than the other.

Clarified Sewage. Sewage from which part or all of the suspended matter has been removed.

Clarity. The transparency of pool water.

Class A, B, C, Roofing. Roof covering materials classified according to their resistance to fire when tested in accordance with ASTM E108; Class A being the highest and Class C the lowest.

Classified Product. A product labeled and listed by an approved laboratory having a factory follow-up and inspection service.

Classroom. A room or building housing a class of students; part of a school.

Classroom Lock. A doorknob in a classroom that has the ability to be locked from the inside.

Classroom Lockset. An assembly mounted in a door in a classroom that contains both a lockable doorknob and deadbolt.

Claw Hammer. A hammer with one end of the head forked for extracting nails.

Clay. 1. A natural mineral aggregate, consisting essentially of hydrous aluminum silicates; it is plastic when sufficiently wetted, rigid when dried, and vitrified when fired to a sufficiently high temperature; used in the manufacture of brick. 2. A heavy soil composed of particles less than 1/256 mm in diameter.

Clay Brick. A type of brick manufactured from fine-grained materials mainly from hydrated silicates of aluminum; soft and cohesive when moist, but becomes hard when baked or fired.

Clay Brick Floor. Hard, baked or fired brick used in flooring applications.

Clay Coping. The use of clay to form a cap or finish on top of a wall, pier, pilaster, or chimney.

Clay Court. A type of surface used for athletic competition, usually tennis or track.

Clay Floor Tile. Quarry tile that is fired and used for flooring.

Clay Pipe. Pipe used for drainage systems and sanitary sewers. made of earthenware and glazed to eliminate porosity.

Clay Tile Partition. An assembly of hollow clay units for constructing interior partitions; the surface is often grooved to receive plaster.

Clay Tile. 1. Earthenware tile that is fired; formed for use on roofs and floors. 2. Quarry Tile.

Clean Out Holes. Openings in first course of one the of a brick wall, or cutouts in face shell of first course of block wall, to enable cleaning out of mortar protrusions and droppings.

Clean out. Removable drainage fitting which permits access to the inside of drainage piping for the purpose of removing obstructions.

Clean. 1. Free of foreign material. 2. In sand or gravel, lack of binder.

Cleaner. 1. Detergent, alkali, acid, or other cleaning material; usually water or steam borne. 2. Solvent for cleaning paint equipment.

Cleaning Masonry. The final removal of excess grout, excess concrete, and construction soil from an exterior masonry structure.

Cleanout. An opening to the bottom of a space of sufficient size and spacing to allow the removal of debris; in plumbing, a fitting in a pipeline which can be easily accessed to remove foreign objects or provide an opening to insert cleaning type devices.

Cleanout Door Cast iron door located at base of chimney to allow access for cleaning the ash pit.

Cleanout Plug. A removable plug in a pipe fitting to enable cleaning a drainage pipe from blockages.

Cleanout, Storm Drain. An opening to a storm drain that allows for the removal of debris.

Clean Room. A dust-free environment that is required for some types of manufacturing, assembly, or fabrication.

Clean Surface. One free of contamination.

Clearance Pocket. In a compressor, a small space in a cylinder from which compressed gas is not completely expelled; also called Clearance Space; for effective operation compressors are designed to have as small a clearance space as possible.

Clearance Space. See Clearance Pocket.

Clear Ceramic Glaze. A colorless or colored transparent ceramic glaze; same as Ceramic Color Glaze except that it is translucent or slightly tinted, with a gloss finish.

Clear Dimension. The dimension between opposing inside faces or walls of an opening or a room; also called Clear Opening.

Clear Glaze. A colorless or colored transparent ceramic glaze; see Clear Ceramic Glaze.

Clearing. An area that has all vegetation and objects removed.

Clear Opening. See Clear Dimension.

Clear Span. The horizontal distance between supports of any structural member.

Clearstory. Clerestory.

Cleat. A small strip or block of material, usually wood, which is fastened to a secure surface and used for attachment or as a toe hold or stopping device for another supporting member.

Cleavage. The natural tendency of certain materials, especially of stones and crystals, to fracture or split in certain definite directions determined by the molecular or physical structure of the material.

Cleavage Membrane. A layer of 15 lb. roofing felt, or an equivalent type of construction paper or polyethylene sheeting, used to isolate a wire reinforced mortar bed for tile from the concrete substrate.

Cleavage Membrane. In tile setting, a membrane such as saturated roofing felt, building paper, or 4 mil polyethylene film installed between the backing and mortar bed to permit independent movement of the tile finish.

Cleft, Natural. See Natural Cleft.

Clerestory. An upper row of windows in a high ceiling room, above the level of a lower adjoining roof; also called clearstory.

Clerk of the Works. Architect's representative on the jobsite; an obsolete term; now called Architect's Project Representative.

Clevis Hanger. A U-shaped metal hanger with the ends drilled to receive a pin or bolt used for attaching or suspending parts or piping.

Clevis Pin. The bolt or pin used to complete the connection in a Clevis Hanger.

CLFMI. Chain Link Fence Manufacturer's Institute.

Clinch. To turn over or flatten the protruding point of a nail to prevent withdrawal.

Clinker. Generally a fused or partly fused by-product of the combustion of coal, but also including lava and portland-cement clinker, and partly vitrified slag and brick.

Clinker Brick. A very hard-burned brick whose shape is distorted or bloated due to nearly complete vitrification.

Clinometer. A device for measuring angles of slope or inclination.

Clip Course. The course of bricks that is resting on a clip joint.

Clip Joint. A brick masonry joint that is thicker than usual to bring that particular course up to a necessary height; the joint should not be any thicker than 1/2 inch.

Clipped Header. In brick masonry, a bat placed to look like a header for purposes of establishing a pattern; also called a false header.

Clips. A classification of devices usually made of wire or sheet metal used to attach various types of lath to supports or to secure adjacent lath sheets.

Clock Receptacle. An electrical outlet box for the connection and support of a clock.

Clock System. A clock which controls the mechanical or electrical devices in a building or facility.

Clock-Timer, Hospital. A clock system in a hospital which can be set at different time gradients to announce, with sound, the passage of a specific amount of time.

Cloister. A covered passage on the side of a court.

Close-Coupled Toilet. A toilet directly connected to its water tank.

Closed Circuit Telephone. See Closed Circuit, 2.

Closed Circuit Television. See Closed Circuit, 2.

Closed Circuit. 1. Electrical circuit in which electrons are flowing. 2. A television or telephone system where the signal is transmitted by wire to a restricted number of receivers.

Closed Container. Container sealed by means of a lid or other device so that neither liquid nor vapor will escape from it at ordinary temperatures.

Closed Joint. In welding, a zero root opening.

Close-Grained Wood. Wood with narrow, inconspicuous annual rings; wood having fine fibers and small closely spaced pores.

Close Nipple. The shortest length of a given size pipe which can be threaded externally from both ends; used to closely connect two internally threaded pipe fittings.

Closer. 1. A Door Closer. 2. The last masonry unit laid in a course. 3. A partial masonry unit used at the corner of a course to adjust the joint spacing, sometimes spelled Closure.

Closet. A storage cabinet or room.

Closet Bend. An elbow drainage fitting connecting a water closet to a branch drain.

Closet, Cedar. See Cedar Closet.

Closet Flange. The fitting attached to a subfloor onto which the toilet bowl is attached.

Closet Pole. A horizontal, round member or rod installed in a closet to provide a place to hang clothes.

Closing Entries. Entries made at the end of the financial period to close all temporary accounts (income and expense) and to transfer the net profit (or loss) to the owner's equity or retained earnings account.

Closure. In brick masonry, supplementary or short length units used at corners or jambs to maintain bond pattern; see Closer, 3.

Closure Strip. A resilient strip, used to close openings created by joining metal panels and flashing.

Closing the Ledger. The operation of closing all income and expense accounts at the end of the period and transferring the net profit (or loss) to the capital or surplus accounts.

Clothes Hanger Rod. See Closet Pole.

Clothing Locker. A cabinet with a lockable door used for storing clothes.

Clothing Presser. A two-part mechanical ironing board that removes wrinkles from clothing with heated, compressive force.

Cloud. 1. A defect in the title to real estate or property; when a property has a cloud on it, it is difficult to sell or complete escrow. 2. A marking on a drawing consisting of a billowing line surrounding portions of the drawing involved in a change, usually identified with a numbered delta symbol.

Cluster Development. A system of close grouping of residential units leaving larger portions of open land around the group.

Clutch, Magnetic. See Magnetic Clutch.

CM. 1. Construction manager. 2. Construction management. 3. Center-Matched.

CMU Grout. Concrete masonry unit grout.

CMU. Concrete Masonry Unit.

CO. 1. Carbon monoxide. 2. Cleanout.

CO_2. Carbon dioxide.

CO_2 Extinguisher. A portable fire-fighting device which dispenses carbon dioxide to extinguish small fires.

CO_2 Indicator. Instrument used to indicate the percentage of carbon dioxide in stack gases.

CO_2 Sprinkler System. An overhead sprinkler system containing Carbon Dioxide, installed and set to turn on when excess heat or smoke activates built-in sensors.

Coagulant. A chemical compound, usually alum, used in swimming pools for the purpose of gathering and precipitating suspended matter.

Coagulate. To change from a liquid into a dense mass; solidify; curdle.

Coagulation. Precipitation of colloids into a single mass; usually caused by excessive heat or catalytic agents.

Coal Hopper Door. Steel door to allow for addition of coal to burning chamber.

Coal Tar. A dark brown to black cementitious material produced by the destructive distillation of bituminous coal.

Coal Tar Bitumen. Coal Tar Pitch.

Coal Tar Epoxy Paint. Paint in which the binder or vehicle is combination of coal tar with epoxy resin.

Coal Tar Pitch. A roofing pitch made from the distillation of bituminous coal; used mainly in dead-level or low-slope roofs; coal-tar pitch comes in a narrow range of softening points from approximately 140° F to 155° F.

Coal Tar Solvent. Derived from the distillation of coal tar.

Coal Tar Urethane Paint. Paint in which the binder or vehicle is a combination of coal tar with polyurethane resin.

Coalesence. In welding, the growing together, or growth into one body, of the base metal parts.

Coarse Aggregate. Concrete aggregate over 1/4-inch diameter; crushed stone, gravel, slag, or other inert materials.

Coarse-Graded Aggregate. One having a continuous grading in sizes of particles from coarse through fine with a predominance of coarse sizes.

Coarse-Grained Wood. Wood with wide conspicuous annual rings in which here is considerable difference between springwood and summerwood; wood with large pores such as oak, ash, chestnet, and walnut; also called Coarse-Textured and Open-Grained.

Coarse Strainer. Basket within a pipeline to trap large debris before the pump.

Coarse-Textured Wood. See Coarse-Grained Wood.

Coat. 1. A thickness, covering or layer of plaster applied in a single operation. 2. See Caot of Paint.

Coated Base Sheet. A felt that has been impregnated and saturated with asphalt and then coated on both sides with harder, more viscous asphalt to increase its impermeability to moisture; a parting agent is incorporated to prevent the material from sticking in the roll.

Coated Roof. A roof, usually flat, that has an asphaltic material applied to it to seal against the elements.

Coating. A layer or covering applied to a surface.

Coating In. Applying a coat of paint.

Coatings. Surface coverings; paints; barriers.

Coat of Paint. One layer of dry paint, resulting from a single wet application; single layer of paint spread at one time and allowed to harden.

Coat Rack. A piece of furniture that is used to hang coats.

Coaxial Cable. A cable composed of two concentric conductors separated by an insulating layer; used for transmitting low voltage electronic signals.

Cobalt Blue. Blue pigment, stable in color; made by heating a mixture of cobalt oxide and aluminum hydrate.

Cobalt Drier. Powerful drier which is soluble in all drying oils; known as surface dryer.

Cobble. 1. To make or put together hastily. 2. A small naturally rounded stone of a size used for paving; a Cobblestone.

Cobblestone. A Cobble.

Cobwebbing. Premature drying of a liquid surface causing a spider web effect.

Cock. A device, as a faucet or valve, to regulate the flow of a liquid.

Cockle. A crease-like wrinkle or small depression in gypsum board face paper usually running in the long or machine direction; also called a wrinkle.

Code. 1. A set of regulations which has been adopted by a governmental unit for the purpose of protecting the public health and safety. 2. The identification marking on the back of sheets of gypsum board; denotes manufacturing plant, date, time, and other details.

Code Blue System. An electronic warning device installed in hospital rooms or health care facility rooms used to notify caregivers of health emergencies in the building.

Code Installation. Refrigeration or air conditioning installation which conforms to the local code and/or the national code for safe and efficient installations.

Code Side. See Back, 3.

Coefficient. 1. A multiplier in any algebraic expression. 2. A multiplier that measures a property of a material or operation.

Coefficient of Conductivity. Measure of the relative rate at which different materials conduct heat; copper is a good conductor of heat and, therefore, has a high coefficient of conductivity.

Coefficient of Cubical Expansion. See Coefficient of Volumetric Expansion.

Coefficient of Expansion. Change in unit length, area or, volume for one degree rise in temperature.

Coefficient of Friction. The mathematical relationship between the weight of an object and the force required to slide it, considering the characteristics of the two materials, the angle of the surfaces, and the angle of the force.

Coefficient of Heat Expansion. The rate of heat loss in BTU per hour through 1 square foot of a wall or other building surface when the difference between the indoor and outdoor air temperatures is 1 degree F; U-Value.

Coefficient of Linear Expansion. The change in unit length for a rise in temperature of 1°F.

Coefficient of Performance (COP). Ratio of work performed or accomplished as compared to the energy used.

Coefficient of Static Friction. The ratio of the limiting friction to the normal pressure (the weight of the moving body).

Coefficient of Thermal Expansion. Change in unit length, area, or volume per degree change of temperature.

Coefficient of Volumetric Expansion. The change in unit volume for a rise in temperature of 1°F; also called Coefficient of Cubical Expansion.

Coffee House. A place serving coffee and other refreshments.

Coffee Room. A room where workers can make and drink coffee, tea, and other refreshments.

Coffee Urn. A device used for the brewing, storage, and serving of coffee, to keep it fresh and hot.

Coffer. A recessed panel in a ceiling or dome.

Cofferdam. A watertight enclosure from which water is pumped to expose the bottom of a body of water and permit construction.

Coffered Ceilings. Ornamental ceilings made up of sunken or recessed panels.

Cog. A gear tooth.

Cohesion. The act or condition of sticking together tightly; tendency to cohere; attractive force between polymers of similar nature which tends to hold them together; property of holding a film together; the soil quality of sticking together.

Cohesive Failure. Rupture of an adhesive joint, such that the separation appears to be within the adhesive.

Cohesive Soil. A soil, such as clay, the particles of which will adhere to one another by means of cohesive and adhesive forces.

Coil. A winding arrangement of a conductor around a core to convert low voltage to high voltage as in a transformer or to create a magnetic field as in a solenoid.

Coil Deck. Insulated horizontal partition between refrigerated space and evaporator space.

Coiled Pipe. Tubing in rows, layers, or windings in steam heating, water heating, refrigeration condensers, and evaporators.

Coin Dryer. A coin-operated public appliance for the drying of clothes.

Coin Washer. A coin-operated public appliance for the washing of clothes.

Cold. The absence of heat; a temperature considerably below normal.

Cold Applied. Roofing products that do not have to be heated before application, unlike tar or asphalt.

Cold Chisel. A hand tool made from a steel bar with a sharpened tip for chipping concrete, stone, and similar materials.

Cold Color. See Cool Color.

Cold Formed Steel. Process of shaping steel without using heat.

Cold Joint Lines. Visible lines on the surfaces of formed concrete indicating the presence of joints where one layer of concrete had hardened before subsequent concrete was placed.

Cold Joint. 1. A visible lineation which forms when the placement of concrete is delayed; the concrete in place hardens prior to the next placement of concrete against it. 2. Any point in a tile installation when tile and setting bed have terminated and the surface has lost its plasticity before work is continued. 3. In road construction, a paving joint in which one strip of asphalt is installed at a different time from the other and bonding is not enhanced.

Cold Junction. That part of a thermoelectric system which absorbs heat as the system operates.

Cold Patch. A roof repair done with cold applied material.

Cold-Pressed Plywood. Interior-type plywood manufactured in a press without external applications of heat.

Cold-Process Roofing. A built-up roof consisting of layers of coated felts bonded with cold-applied asphalt roof cement and surfaced with a cutback or emulsified asphalt roof coating.

Cold Rolled Steel. Steel rolled to its final form at a temperature at which it is no longer plastic.

Cold-Setting Resin Glue. A resin-base glue that comes in powder form and is mixed with water.

Cold Wall. Refrigerator construction which has the inner lining of refrigerator serving as the cooling surface.

Cold Water Paint. The paint in which the binder or vehicle portion is composed of casein, glue or a similar material dissolved in water; usually employed on concrete, masonry or plaster surfaces.

Cold Worked Steel. Steel formed at a temperature at which it is no longer plastic, as by rolling or forging.

Cold-Checking. Checking caused by low temperature.

Coliseum. A large sports stadium or building designed like the Roman Coliseum for public entertainments.

Collage. An artistic composition in which various materials such as photographs, pieces of fabric, metals, and wood are arranged and glued to a backing.

Collapse. To cave or fall in or give way; failure of a structure.

Collar. A compression ring around a small circular opening.

Collar Beam. A horizontal beam near the top of a trussed rafter system attached to opposing rafters to resist their spreading; also called a Collar Tie or a Collar Beam.

Collar Joint. The vertical mortar joint between the masonry.

Collar Tie. See Collar Beam.

Collateral. Property pledged or in the possession of a creditor to guarantee payment of an obligation by a debtor.

Collector. 1. In structural analysis, a force transfer element that collects loads from a diaphragm (horizontal element) and transfers them to the shear walls (vertical element). 2. Any of a wide variety of devices (flatplate, concentrating, etc.) used to collect solar energy and convert it to heat; a Solar Collector.

Collector Efficiency. The ratio of heat energy extracted from a collector to the quantity of solar energy striking the cover expressed in percent.

Collector Elements. Elements that serve to transmit the inertial forces with the diaphragms to members of the lateral-force-resisting systems.

Collector Streets. Connecting roads between arterial streets, not necessarily continuous, to discourage through traffic; intersections often controlled by traffic lights; parking permitted under various conditions; pedestrian crossing controlled.

Collet. A metal band, flange, or ferrule.

Colloid. A mixture containing ultramicroscopic particles of one substance scattered evenly throughout another; also known as a dispersion.

Colloidal Suspension. A substance divided into fine particles which remains in permanent suspension in a liquid.

Collusion. A secret combination or conspiracy between two or more persons having a fraudulent or deceitful purpose; the illegal practice of contractors agreeing to submit higher bids so that a chosen contractor's bid will be low.

Colonial Base. An architectural style of a board or molding used against the bottom of walls to cover the joint with the floor and to protect the walls from dents and scuffs.

Colonial Molding. A style of molding from eighteenth century English Georgian, reproduced and incorporated into buildings in America.

Colonnade. A set of columns occurring at regular intervals.

Color (Colour). The visual appearance of objects and light sources in terms of hue, lightness, and saturation for objects and hue, brightness, and saturation for light sources; a hue as contrasted with black, white, and gray.

Color, Complementary. See Complementary Color.

Colored Finishes. Plaster finish coats containing integrally mixed color pigments or colored aggregates.

Colored Grout. Commercially prepared grout consisting of carefully graded aggregate, portland cement, water dispersing agents, plasticizers, and color fast pigments.

Color-Fast. Non fading.

Colorimeter. See Chromometer.

Coloring Strength. The relative strength or ability of pigments to color base material which is white or light in color.

Color-In-Japan. A paste formed by mixing a color pigment with Japan drier; used principally for tinting.

Color-In-Oil. A paste formed by mixing a color pigment in linseed or other vegetable oil; used principally for tinting.

Color Man. In painting, the individual, either the journeyman or contractor, who is an expert in tinting and matching colors.

Color Pigments. Pigments such as blue, red, etc. which absorb a portion of the light which falls upon them and reflect or return to the eye certain groups of light bands which enable us to recognize various colors.

Color, Primary. See Primary Color.

Color Retention. When a paint product exposed to the elements shows no signs of changing color it is said to have good color retention.

Colossus. A statue of colossal size or proportions.

Columbarium. A structure containing vaults for cinerary urns.

Column. 1. A structural member used primarily to support axial compression loads and with a height of at least three times its least lateral dimension; an upright structural member acting primarily in compression. 2. A square, rectangular, or cylindrical support for roofs, ceilings, and so forth, composed of base, shaft, and capital.

Column Base. 1. The part which forms the bottom of a vertical supporting pillar. 2. The plate beneath a column that distributes the load.

Column Base Plate. The part of a structure on which the column base is set.

Column Cage. As assembly of vertical reinforcing bars and ties for a concrete column.

Column Cap. See Capital, 2.

Column Capital. The uppermost member of a column crowning the shaft and taking the weight of the beam or girder.

Column, Concrete. A long, relatively slender, supporting pillar made from concrete and reinforcing steel.

Column Cover. A structure that forms the uppermost part of a column.

Column Fireproofing. The act or process of applying fire-retardant materials to a column.

Column Footing. Concrete support for a column; commonly known as individual footing, generally square or rectangular in shape.

Column Formwork. The mold or sheathing that forms the outline of a column into which the concrete is placed or poured.

Column Furring. Strips of wood or metal applied to a column to provide a fastening surface for a finish covering.

Columniation. The arrangement of columns in a building.

Column Pier. A foundation member of plain or reinforced concrete to support a column.

Column, Precast. A column that has been cast and cured in other than its final position.

Column Reinforcing. The embedded steel bars to strengthen a concrete column.

Column, Sheetrock. A column made from gypsum wallboard.

Column Spiral. A continuous coil of steel reinforcing used in a concrete column.

Column Strip. The zone of a two-way concrete floor or roof structure that is centered on a line of columns.

Column Tie. A single loop of steel bar, usually bent into a rectangular configuration, used to tie the reinforcing cage in a concrete column.

Column, Timber. Structural lumber, five inches or more in its least dimension used as a vertical compression member.

Column, Wood. Vertical wood structural member, usually carrying a beam.

Comb Board. See Saddle Board.

Combination Column. A column in which a structural steel member, designed to carry the principal part of the load, is encased in concrete which carries the remainder of the load.

Combination Doors or Windows. Doors or windows with self-storing or removable glass and screen inserts; the need for handling a different unit each season is thus eliminated.

Combination Frame. In light wood framing, a combination of the principal features of the full and balloon frames.

Combination Receptacle. An electrical fixture with an outlet for plugging in an electrical appliance, along with an electric switch for operating another circuit.

Combination Switch. A switch that includes a switch and a receptacle.

Combined Chlorine. Chlorine combined with other substances; though available to disinfect pool water; chlorine in this form is less effective than free chlorine.

Combined Footing. A concrete footing which supports two or more columns.

Combined Stress. The combination of axial and bending stresses in a structural member.

Combined Water. The water, chemically held as water of crystallization, by the calcium sulphate dihydrate, or hemihydrate crystal.

Combustible. Capable of being burned.

Combustion. Burning; consumption by fire; the development of light and heat from the chemical combination of a substance with oxygen.

Comfort Chart. Chart used in air conditioning to show the dry bulb temperature, humidity and air movement for human comfort conditions.

Comfort Cooler. System used to reduce the temperature in the living space in homes; these systems are not complete air conditioners as they do not provide complete control of heating, humidifying, dehumidification, and air circulation.

Comfort Zone. Area on psychrometric chart which shows conditions of temperature, humidity, and sometimes air movement in which most people are comfortable.

Commerce. 1. The exchange or buying and selling of commodities, especially on a large scale involving transport within a country or between countries. 2. The principles and techniques of business and office systems.

Commercial Carpet. Carpet that is highly resistant to heavy traffic.

Commercial Door. A type of door used for strength and durability in commercial building applications.

Commercial Facilities. Facilities that are intended for nonresidential use and whose operations will affect commerce, including factories, warehouses, office buildings, and other buildings in which employment may occur.

Commercially Acceptable Standards. Of workmanship, the standard of work produced by the average competent craftworker, allowing a reasonable amount of imperfections; not perfect.

Commercial Matching. Matching of colors within acceptable tolerances, or with a color variation that is barely detectable to the naked eye.

Commingling of Funds. To combine funds belonging to different accounts into a single account

Commission. 1. The professional fee paid to a broker or agent for services, usually calculated as an agreed percentage of the sales or lease consideration. 2. A fee paid to an agent or employee for transacting a piece of business or performing a service.

Commissioning. Start-up of a building that includes testing and adjusting HVAC, electrical, plumbing, and other systems to assure proper functioning and adherence to design criteria; also includes the instruction of building representatives in the use of the building systems.

Commitment. 1. A legally enforceable agreement to do something in the future. 2. A pledge by a lending institution to make a real estate loan under certain stated conditions.

Commode. 1. A chest of drawers. 2. A chamber pot concealed in a chair with a hinged cover.

Common Bolt. An ordinary carbon steel bolt.

Common Bond. Brickwork laid with each five courses of alternating stretchers followed by one course of headers.

Common Brick. Inexpensive brick, not selected for appearance, that is used as filler or backing. See Building Brick.

Common Enemy Doctrine. The legal doctrine that flood waters are a common enemy and that property owners may fight to protect their property regardless of the damage to neighboring property.

Common Rafter. A rafter that is square with the plate and extends to the ridge.

Common Seal-P Trap. A P-trap with a water seal depth of 2 to 4 inches.

Common Vent. A vent that connects at the junction of two fixtures, acting as a vent for both fixtures.

Common Waster Pipe. Used when there are two sets of bathroom fixtures on opposite sides of a wall with their drain connections opposite each other; the fixtures may drain into the same waste and also have dual vents; fixtures which are directly across from each other may have common wastes and dual vents.

Communication. An act or instance of transmitting information.

Communication Cable. A cable fortransmission of telephone, television, and computer signals.

Communication Circuit. A circuit which is part of a central station system; such circuits include telephone, telegraph, district messenger, inter-communications, public address systems, fire and burglar alarms, watchmen and sprinkler supervisory circuits with their associated operating and signaling power supply equipment.

Communication Manhole. Any of various types of manholes used in the installation of communication conduits or cables.

Community Kitchen. A room in a hotel or lodging house used or intended to be used by the occupants of two or more guest rooms for cooking or preparing food.

Community Shopping Center. An intermediate size of Shopping Center.

Commutator. Part of rotor in electric motor which conveys electric current to rotor windings.

Compact Borrow. Fill acquired from excavation that has been compacted.

Compacted Concrete. Freshly poured concrete that has been packed tighter by vibration, tamping, or a combination of both to remove voids.

Compacted Yards. Cubic measurement of soil or rock after it has been placed and compacted in a fill.

Compaction. The process whereby the volume of freshly placed material is reduced or flattened by vibration or tamping, or some combination of these; reduction in bulk of fill by rolling, tamping or soaking; insufficient compaction of asphalt pavement courses may result in channeling on the pavement surface.

Compaction Tile. A hard tile surfacing unit made from a mixture of chemicals; the finished surface can be the mixture of chemicals or can be marble chips to create a terrazzo finish; the unit is made hard by the set of the chemicals and the product is not fired as in the manufacture of ceramic tile.

Compactor. 1. A mechanical device that compresses objects into smaller units. 2. A machine in an industrial setting that compacts or compresses materials. 3. A machine in a kitchen that compresses or compacts materials by using hydraulic weight, force or vibration.

Companion Flange. A pipe connection device that is machined and drilled to match another flange on a pipe or fitting.

Comparative Negligence. The legal doctrine that wrongdoers should pay damages proportional to their fault.

Compartment Kiln. A kiln in which the total charge of lumber is dried as a single unit; it is designed so that, at any given time, the temperature and relative humidity are essentially uniform throughout the kiln; the temperature is increased as drying progresses, and the relative humidity is adjusted to the needs of the lumber.

Compartment, Shower. See Shower Compartment.

Compass. 1. A drafting tool; an adjustable device for drawing circles; similar to a dividers but with a point on one leg and a pen or pencil on the other; often called compasses or a pair of compasses. 2. A Magnetic Compass.

Compatibility. Ability to mix with or adhere properly to other components or substances.

Compensating Errors. Errors that are small in magnitude and in which the pluses and minuses (overs and unders) tend to offset each other; compare with Cumulative Errors.

Compensatory Damages. An amount calculated to compensate a party for economic loss caused by the wrongful act of another.

Competitive Bidding. The process of two or more contractors submitting proposals for the same work at the same time.

Complementary Color. A color that combined with its complement yields gray; complementary pairs include red-green, yellow-violet, and blue-orange.

Complete Fusion. In welding, fusion which has occurred over the entire base- metal surfaces exposed for welding.

Completion Bond. A bond guaranteeing to the lender that the project will be completed free of liens.

Completion Date. The date stipulated in the construction contract for substantial completion.

Completion of Construction. When the work of the construction contract is complete.

Component. 1. A constituent part or ingredient. 2. One of two or more forces which, acting together, have the same effect as a single force, called the resultant.

Component Depreciation. A method of computing depreciation of property by its individual parts rather than as a whole.

Composite. An order of Greek architecture consisting of elements of the Ionic and the Corinthian orders.

Composite Beam. A beam that is composed of two different materials; for example, a wood and steel beam, or a steel beam and concrete slab, in which the two act as one.

Composite Board. A board that is made of several compressed materials; used for sheathing, wallboard, or as an insulation or acoustical barrier.

Composite Column. A column in which a steel or cast-iron structural member is completely encased in concrete containing spiral and longitudinal reinforcement.

Composite Concrete Flexural Composition. A precast concrete member and cast-in-place reinforced concrete so interconnected that the component elements act together as a flexural member.

Composite Decking. A type of decking construction made up of different materials, such as concrete, wood, and steel.

Composite Metal Decking. Corrugated steel decking manufactured in such a way that it bonds securely to the concrete floor fill to form a reinforced concrete deck.

Composite Steel Deck Shoring. Shoring used on composite steel decks during the placement of concrete.

Composite Wall. A masonry wall that incorporates two or more different types of masonry units, such as clay bricks and concrete blocks.

Composition. 1. The factors or parts of any substance or material. 2. The quantitative or qualitative makeup of any chemical.

Composition Roofing. Any asphaltic roofing.

Composition Shingles. Shingles made or formed from composition roofing material.

Compound Gauge. Instrument for measuring pressures both above and below atmospheric pressure.

Compound Refrigerating Systems. System which has several compressors or compressor cylinders in series; used to pump low-pressure vapors to condensing pressures.

Compreg. Wood in which the cell walls have been impregnated with synthetic resin and compressed; this process reduces swelling and shrinking and increases density and strength; compare with Impreg.

Comprehensive General Liability. Insurance policy covering comprehensive general liability.

Compression. 1. Force which tends to crush adjacent particles of a material together and cause overall shortening in the direction of its action; stress which tends to shorten a member. 2. The increase of pressure on a fluid by using mechanical energy.

Compression Bars. Steel used to resist compression forces.

Compression Block. An extremely short wooden column.

Compression Connector. A connecting device which when attaching uses a force that pushes together and squeezes.

Compression Coupling. A connecting device which creates a force that pushes together and crushes.

Compression Elbow. A connecting device which joins two pipes at 90 degrees which uses a force that pushes together and crushes.

Compression Failure. Deformation of the wood fibers resulting from excessive compression along the grain in direct end compression or in bending; it may develop in standing trees due to bending by wind or snow or to internal longitudinal stresses developed in growth, or it may result from stresses imposed after the tree is cut.

Compression Faucet. One designed to stop the flow of water by the action of a flat washer closing against a seat.

Compression Fitting. Bends, couplings, crosses, elbows, tees, and unions which use a force when connecting that pushes together and squeezes a metal or rubber gasket.

Compression Gasket. A synthetic or rubber strip or washer that seals by being squeezed tightly.

Compression Gauge. An instrument used to measure positive pressures, those above atmospheric pressures, only; these gauges are usually calibrated from 0 to 300 psig (0-21.1 kg./cm^2).

Compression Lug. A connector for fastening the end of a wire to a terminal that uses a force that pushes together and crushes.

Compression Ratio. In a refrigeration compressor, the ratio of the volume of the clearance space to the total volume of the cylinder, in refrigeration it is also used as the ratio of the absolute low-side pressure to the absolute high-side pressure.

Compression Reinforcement. Steel reinforcing bars in a reinforced beam that are placed near the top of the beam to assist in resisting compressive forces.

Compression Ring. A structural element, circular in plan, which is in compression because of the action of the rest of the structure; for a dome, it would occur as a collar around a circular opening, or oculus, at the top; for a suspended roof, it would be around the outside edge.

Compression Splice. A connection of two similar materials by a force which tends to shorten a member.

Compression Strength. The ability of a structural material to withstand compression forces; the measured maximum resistance of a concrete or mortar specimen to axial loading; expressed as force per unit cross-sectional area; or the specified resistance used in design calculations, in the U.S. customary units of measure expressed in pounds per square inch.

Compression Valve. One designed to stop the flow of water by the action of a flat washer closing against a seat.

Compression Wood. Abnormal wood formed on the lower side of branches and inclined trunks of softwood trees; compression wood shrinks excessively lengthwise, as compared with normal wood.

Compressive Strength. The measured resistance of a concrete or mortar specimen to axial loading expressed as pounds per square inch of cross-sectional area; the maximum compressive stress which material, such as portland cement, concrete, or grout is capable of sustaining.

Compressor. A machine that compresses gases or air; pump of a refrigerating mechanism which draws a low pressure on the cooling side of the refrigerant cycle and squeezes or compresses the gas into the high-pressure or condensing side of the cycle.

Compressor Displacement. Volume, in cubic inches, represented by the area of the compressor piston head or heads multiplied by the length of the stroke.

Compressor, External Type. See Compressor, Open Type.

Compressor, Hermetic. See Hermetic Compressor.

Compressor, Multiple Stage. Compressor having two or more compressive steps. Discharge from each step is the intake pressure of the next in series.

Compressor, Open Type. Compressor in which the crankshaft extends through the crankcase and is driven by an outside motor; also called External Drive Compressor.

Compressor, Reciprocating. See Reciprocating Compressor.

Compressor, Rotary. Compressor which uses vanes, eccentric mechanisms or other rotating devices to provide pumping action.

Compressor Seal. Leakproof seal between crankshaft and compressor body in open type compressors.

Compressor, Single Stage. Compressor having only one compressive step between low-side pressure and high-side pressure.

Computer Cable. Coaxial cable which transmits computer signals.

Computerese. Arcane jargon used by computer technologists.

Computer Floor. Special flooring designed to prevent electrostatic buildup and sparking in a computer room; usually elevated above the existing floor, to facilitate the running of wires between the components in the room.

Computer Hardware. All of the physical equipment, including the input units (keyboard and mouse), processing unit, and output units (screen, and printer).

Computer Language. A code used to write a computer program.

Computer-Room AC. An air-conditioning unit used in a computer room.

Computer Software. The programs that make a computer function.

Concave. Curved or rounded inward like the inside of a bowl; the opposite of convex.

Concave Fillet Weld. A fillet weld having a concave face.

Concave Joint. In masonry, a mortar joint formed with a special tool or a bent iron rod; this joint is weather resistive and inexpensive.

Concavity. The maximum distance from the face of a concave fillet weld perpendicular to a line joining the toes.

Concealed. Rendered inaccessible by the structure or finish of the building.

Concealed Grid. A suspended ceiling framework that is completely hidden by the tiles or panels it supports.

Concealed Picture Mold. A recessed horizontal slot in a plaster wall, formed by a sheet metal screed, used to hang pictures and other objects.

Concealed Z Bar. A hidden z-shaped bar that is used as a wall tie.

Concentrated Load. A load which acts at one point or small area of a structure or member.

Concentrating Collector. A device that uses reflective surfaces to concentrate the sun's rays onto a smaller area, where they are absorbed and converted to heat energy.

Concentration. 1.The strength of a solution. 2. A concentrated mass or thing.

Concentric. Having a common center or axis.

Concrete. A mixture of portland cement, fine aggregate, coarse aggregate, admixtures, air, and water.

Concrete Accessory. An implement or device used in the formwork, pouring, spreading, or finishing of concrete surfaces.

Concrete Admixture. A substance added to concrete to aid in imparting color, control workability, help in waterproofing, control setting, and to entrain air.

Concrete Beam. A horizontal structural member which transversely supports a load and transfers the load to vertical members, made of a composite material consisting of sand, coarse aggregate, cement and water.

Concrete Block Bar Supports. Precast concrete blocks, with or without tie wires used to support bars above the sub-grade or to space bars off vertical forms and above horizontal forms.

Concrete Block Removal. The act or process of demolition of a concrete block structure.

Concrete Block. A hollow concrete masonry unit made from portland cement and suitable aggregates such as sand, gravely crushed stone, bituminous or anthracite cinders, burned clay or shale, pumice, volcanic scoria, air-cooled or expanded blast furnace slags with or without the inclusion of other materials.

Concrete Brick. A solid concrete masonry unit the same size and proportion as a clay brick.

Concrete Burlap. A curing concrete surface that has had a coarse fabric of jute, hemp, or less commonly, flax applied, for use as a water-retaining covering.

Concrete, Cellular. See Cellular Concrete.

Concrete Column. A long, relatively slender, supporting pillar made from concrete and reinforcing steel.

Concrete Cover. The distance from a reinforcing bar to the outside of a concrete member; also referred to as fireproofing, clearance, or concrete protection.

Concrete Cutting. Scoring or cutting of concrete or masonry with a saw equipped with a carborundum blade; commonly done by specialty subcontractors with customized equipment.

Concrete Dowel. A pin of reinforcing steel embedded in concrete to strengthen two pieces where they join or to create a place where other pieces can be fastened to it.

Concrete, Fibrous. See Fibrous Concrete.

Concrete, Field. See Field Concrete.

Concrete Filled Pile. A long slender construction element filled with concrete, driven in the ground for the purpose of supporting a load.

Concrete Finish, Board. Wooden boards placed in the concrete formwork as for liners to provide a wood-pattern finish to the completed reinforced concrete.

Concrete Finish. The act or process of the final compaction and finishing operations of curing concrete.

Concrete, Foamed. See Foamed Concrete

Concrete, Green. See Green Concrete

Concrete Headwall. The end of a culvert or drain constructed of concrete.

Concrete, Lightweight. Concrete that has substantially lower weight than that made from gravel or crushed stone.

Concrete Manhole. A vertical access shaft from the surface to the underground, constructed of concrete.

Concrete Masonry Unit (CMU). See Concrete Block.

Concrete Mat. A grid of metal reinforcement for concrete foundations, slabs, or mats.

Concrete Mix. The amount of each material specified, portland cement, fine aggregate, coarse aggregate, admixture, and water.

Concrete Nail. A hardened steel nail that may be driven into concrete or masonry.

Concrete, Normal Weight. See Normal Weight Concrete.

Concrete Parking Barrier. A concrete structure placed to act as a barrier against vehicular encroachment.

Concrete Paving. The use of concrete to make a hard surface in areas such as walks, roadways, ramps, and parking areas.

Concrete Pile. A precast slender reinforced concrete member that is embedded in the soil, by driving or inserting into a predrilled hole.

Concrete Pipe. Pipe manufactured from concrete; the manufacturing is done in a plant under controlled conditions; usually used for drainage but may also be used for sanitary sewers.

Concrete Placement. The placing and finishing of concrete during a continuous operation; also known as pouring.

Concrete, Plain. See Plain Concrete.

Concrete Plank. A solid or hollow-core, flat-beam used for floor or roof decking; usually precast and prestressed.

Concrete Planter. A concrete reservoir to hold soil for plantings.

Concrete Pole. A vertical member made of concrete.

Concrete Post. A vertical structure made of concrete which carries stresses in compression.

Concrete, Precast. Concrete parts that are cast on- or off-site and, after curing and hardening, are installed in their final position of use.

Concrete, Prestressed. See Prestressed Concrete.

Concrete Pump. An apparatus which forces concrete to the placing position through a pipeline or hose.

Concrete, Pumped. See Pumped Concrete.

Concrete Receptor. A precast concrete structure which forms the drain area of a shower.

Concrete, Reinforced. See Reinforced Concrete.

Concrete Reinforcement. Steel rods that are embedded in wet concrete to give additional strength.

Concrete Removal. The act or process of demolition of old concrete into manageable pieces.

Concrete Repair. To restore concrete by replacing a section or repairing what is broken.

Concrete Restoration. The rebuilding of the surface of concrete to approach as nearly as possible the original form.

Concrete Saw Cut. A cut in hardened concrete utilizing diamond or silicone-carbide blades or discs.

Concrete Sheet Piling. A row of concrete piles driven in close contact to provide a tight wall to resist the lateral pressure of water, adjacent earth or other materials.

Concrete Sleeper. Strips of wood placed on a rough concrete floor to which the finished wood floor is nailed.

Concrete, Structural Lightweight. Concrete containing lightweight aggregate, not exceeding 115 pounds per cubic foot.

Concrete, Terrazzo. See Terrazzo.

Concrete Testing. Testing to determine the plasticity or strength of concrete.

Concrete Topping. A rich mixture of fine aggregate concrete used to top concrete floor surfaces for durability, safety and appearance.

Concrete Waterproofing. An act or process of adding a material to concrete to make it impervious to water or dampness.

Concurrent. The point at which the line of action of several forces meet.

Concurrent Heating. The application of supplemental heat to a structure during a welding or cutting operation.

Condemnation. A procedure by which private property is acquired for public use by the legal process of eminent domain.

Condensate. A product obtained by cooling vapors of a substance being distilled.

Condensate Pump. A pump that removes water or condensation from an air-cooling unit.

Condensation. The process of changing from a gaseous to a liquid state, especially as applied to water; liquid drops which form when a vapor is chilled below its boiling point; water droplets that deposit on surfaces whose temperature is below the dew point.

Condensation Drain. A drain pipe to carry off water condensed from the atmosphere.

Condense. Action of changing a gas or vapor to a liquid.

Condenser. A heat exchanger in a refrigeration cycle used to discharge heat to the outside; commonly used types are water cooled, air cooled, and evaporative; condenser water is normally circulated through a cooling tower through which heat is distributed to the atmosphere.

Condenser, Air-Cooled. Heat exchanger which transfers heat to surrounding air.

Condenser Comb. Comb-like device, metal or plastic, used to straighten the metal fins on condensers or evaporators.

Condenser Fan. Forced air device used to move air through air-cooled condenser.

Condenser, Water Cooled. Heat exchanger designed to transfer heat from hot gaseous refrigerant to water.

Condensing Pressure. Pressure inside a condenser at which refrigerant vapor gives up its latent heat of vaporization and becomes a liquid; this varies with the temperature.

Condensing Temperature. Temperature inside a condenser at which refrigerant vapor gives up its latent heat of vaporization and becomes a liquid; this varies with the pressure.

Condensing Unit Service Valves. Shutoff valves mounted on condensing unit to enable service technicians to install or service unit.

Condensing Unit. Part of a refrigerating mechanism which pumps vaporized refrigerant from the evaporator, compresses it, liquefies it in the condenser and returns it to the refrigerant control.

Conditioned Air. Air that has been heated, cooled, humidified, or dehumidified to maintain an interior space within the comfort zone; also called Tempered Air.

Conditioned Water. In swimming pools, water treated with cyanuric acid or chlorinated isocyanurate to prevent chlorine from dissipating in sunlight.

Condominium. Individual ownership of a unit in a multi-unit development, as dwellings, offices, storage, or manufacturing spaces; also includes ownership of an interest, in common with other owners, of common areas and facilities that serve the structure.

Conductance. 1. The property of a material to conduct electric current. 2. A property of a material to conduct heat; a property of a slab of material equal to the quantity of heat, in BTU. per hour, that flows through 1 square foot of the slab, when a 1 degree F. temperature difference is maintained between the two sides.

Conduction. Heat transfer from one particle to another in direct contact with it.

Conductive. Having the quality or power of conducting or transmitting heat, electricity, or static electricity.

Conductive Floor. Flooring material specifically designed to prevent electrostatic buildup and sparking.

Conductive Mortar. A tile mortar to which specific electrical conductivity is imparted through the use of conductive additives.

Conductive Terrazzo. A type of tile that prevents electrostatic buildup and sparking, that is made from specific conductive materials.

Conductive Tile. Tile made from special body compositions or by methods that result in specific properties of electrical conductivity while retaining other normal physical properties of ceramic tile.

Conductivity (k). Ability of a substance to conduct or transmit heat or electricity; the reciprocal of resistivity, 1/r.

Conductor. A substance or body capable of transmitting electricity, heat or sound.

Conductor, Aluminum. An aluminum wire or cable for transmitting electrical current.

Conductor, Copper. A copper wire or cable for transmitting electrical current.

Conductors. 1. Pipes for conducting water from a roof to the ground or to a receptacle or drain; downspout. 2. Any electrical wire used to convey electricity.

Conductor, Stranded. A number of fine wires twisted around a center wire or core, used as a single electric conductor.

Conduit. A protective sleeve or pipe commonly used for individual electrical conductors.

Conduit, Aluminum. A pipe constructed of a light alloy material used to enclose electric wires to protect them from damage.

Conduit Bushing. A threaded metal or plastic pipe connector used to connect conduit to a box or other housing where the hole is not threaded.

Conduit Cap. A cap placed on the end of a length of conduit to protect the threads or terminate the conduit run.

Conduit, EMT. See EMT Conduit.

Conduit, Flexible. See Flexible Conduit.

Conduit, Galvanized. See Galvanized Conduit.

Conduit Locknut. A threaded connector used where conduit enters an electrical box or other housing where the hole in not threaded; the locknut is screwed onto the threaded end of the conduit on the outside of the box and connected to a bushing on the inside of the box until secure.

Conduit Locknut. A threaded connector used where conduit enters an electrical box or other housing where the hole is not threaded; the locknut is screwed onto the threaded end of the conduit on the outside of the box and connected to a bushing on the inside of the box until secure.

Conduit, Plastic Coated. Asee Plastic Coated Conduit.

Conduit Plug. A fitting that is screwed into the ends of conduit or conduit fittings.

Conduit, PVC. See PVC Conduit.

Conduit, Steel. See Steel Conduit.

Cone. 1. A solid figure with a circular (or other curved) plane base, tapering up to a point. 2. The dry fruit of a conifer; a pine cone. 3. The conical part of a gas flame next to the orifice of the tip.

Cone, Slump. See Slump Cone.

Conference Room. A room for meetings.

Configuration. An arrangement of parts or elements in a particular form or figure.

Congo Gum. A gum resin obtained from the Congo region of Africa.

Conifer. Any evergreen of a group usually bearing cones; conifers produce softwood lumber.

Connection. The union, or joint, of two or more distinct elements.

Connection, Bus Duct. See Bus Duct Connection.

Connector, Compression. See Compression Connector.

Connector, Die Cast. A connecting device that has been cut, formed and threaded by die tools.

Connector Set Screw. A screw on a connector fitting that when tightened connects two components together.

Consent of Surety. A written consent of the surety to pay the final payment to the contractor, reduce or eliminate the retainage, or make any change in the contract conditions.

Conservation. The careful preservation and protection of something, such as a building or the environment.

Conservatism. A principle of accounting, the object of which is to place each asset item on the balance sheet at a conservative figure or low end of its price range.

Conservatory. A greenhouse for tending and displaying plants.

Consideration. A benefit (or money) coming from a promisee to a promisor in exchange for the promisor's agreement to perform an act.

Consistency. 1. The fluidity or viscosity of a liquid or paste; resistance of a product to flow. 2. The degree of plasticity of fresh concrete or mortar; the normal measure of consistency is slump for concrete and flow for mortar. 3. The degree of fluidity or plasticity of asphalt cement at any particular temperature; the consistency of asphalt cement varies with temperature; therefore, it is necessary to use a common or standard temperature when comparing the consistency of one asphalt cement with another, the standard test temperature being 140° F (60° C).

Consistency, Normal. Of gypsum plaster or gypsum concrete, see Normal Consistency.

Console. 1. An ornamental bracket supporting a shelf. 2. A cabinet or panel containing controls and switches for operating lighting, sound, television, or radio equipment.

Consolidation. Process of working fresh concrete so that a closer arrangement of particles is created and the number of voids is decreased or eliminated.

Consolidation Compaction. This is usually accomplished by vibration of newly placed concrete to its minimum practical volume, to mold it within form shapes and around embedded parts and reinforcement and to eliminate voids other than entrained air.

Consortium. A combination of companies organized to invest in an enterprise usually too large for them individually.

Constant. A component of a relationship between variables that does not change its value.

Constant Air Volume System. Air handling system that provides a constant air flow while varying the temperature to meet heating and cooling needs.

Constrictor. Tube or orifice used to restrict flow of a gas or a liquid.

Construct. Make by putting materials or parts together; build; erect.

Constructibility. The capacity of a certain design of being developed into construction.

Construction. 1. The act or process of building. 2. Buildings and structures that have been built. 3. The method by which the carpet is made (loom or machine type) and other identifying characteristics, including pile rows per inch, pitch, wire height, number of shots, yarn count and plies, pile yarn weight, and density.

Construction Change Directive. A document that directs a change in the work of the project, prepared by the architect and signed by the owner and architect. This change can adjust the contract sum and/or contract time. This document may be used in the absence of the contractor's agreement on the terms of a change order.

Construction Documents Phase. One of the standard phases of architectural service (Schematic Design Phase, Design Development Phase, Construction Documents Phase, Bidding or Negotiation Phase, and Construction Phase-Administration of the Construction Contract).

Construction Documents. The drawings and specifications that describe the construction requirements.

Construction Industry Arbitration Rules. Arbitration rules administrated by the AAA and referred to in AIA's standard agreements for construction and architectural services.

Construction Joint. The contact between the placed concrete and concrete surfaces, against or upon which concrete is to be placed and to which new concrete is to adhere, that has become so rigid that the new concrete cannot be incorporated integrally by vibration with that previously placed.

Construction Lender. An institution or individual that lends money to a borrower for the purpose of constructing improvements on property.

Construction Lime. An industrial form of calcium oxide that is added to give it cling or adhesion.

Construction Loan. A short term interim loan to pay for the construction of a building.

Construction Management. Activities over and above normal architectural and engineering services, conducted during the predesign, design, and construction phases, that contribute to the control of time and cost.

Construction Manager. A person or entity who provides construction management services, either as an advisor or as a contractor.

Construction Phase-Administration of the Construction Contract. One of the standard phases of architectural service (Schematic

Design Phase, Design Development Phase, Construction Documents Phase, Bidding or Negotiation Phase, and Construction Phase-Administration of the Construction Contract).

Construction Schedule. A time table setting out the times for starting and completing each of the operations required for the construction of a building or other project.

Constructive Notice. The giving of notice by recording it in the office of the county recorder or placing it in a newspaper.

Consultant. One who provides services as an advisor.

Consumable Insert. See Backing Filler Metal.

Contact Ceiling. A ceiling which is secured in direct contact with the construction above without use of furring.

Contact Cement. Neoprene rubber-based adhesive which bonds instantly upon contact of parts being fastened.

Contact Fireproofing. See Fireproofing.

Contact Splice. A means of connecting reinforcing bars by lapping in direct contact.

Contaminant. 1. A substance that taints, infects, or pollutes; an impurity. 2. In a refrigeration system, substances such as dirt, moisture, or other matter foreign to refrigerant or refrigerant oil in system.

Contaminated. 1. No longer as originally manufactured or no longer pure because of contact with some foreign substance. 2. In tile work, stained tile as a result of carton and tile being saturated by moisture, oils, solvents or other materials.

Contingency Allowance. A sum in a construction budget to cover unforeseen expenses.

Continuity. 1. Uninterrupted electrical path in a circuit. 2. Degree of being intact or pore free.

Continuous Beam. A beam that is supported by more than two supports.

Continuous Cycle Absorption System. System which has a continuous flow of energy input.

Continuous Duty. In an electrical system, a requirement of service that demands operation at a substantially constant load for an indefinitely long time.

Continuous Footing. A concrete footing that supports a wall or two or more columns; the footing can vary in width and depth, also called a Strip Footing.

Continuous High Chairs. Welded wire bar supports consisting of a top longitudinal supporting wire with evenly spaced legs welded thereto and used to support bars near the top of slabs.

Continuous Hinge. A hinge designed to be the same length as the moving part (cabinet door or lid) to which it is applied; also called Piano Hinge.

Continuous Load. An electrical load where the maximum current is expected to continue for three hours or more during any one period of time.

Continuous Mix Plant. A manufacturing facility for producing asphalt paving mixtures that proportions those aggregate and asphalt constituents into the mix by a continuous volumetric proportioning system without definite batch intervals.

Continuous Ridge Vent. A screened, water-shielded ventilation opening that runs continuously along the ridge of a gable roof.

Continuous Vent. A vertical vent pipe that is a continuation of the waste drain.

Continuous Waste and Vent. A vertical vent pipe that is in a continuous line with the waste line; below the point where the fixture drains into the pipe is the waste line; above this point is the vent.

Continuous Waste. Two or more fixtures that use a single trap.

Contour Interval. The difference in vertical measurement between adjacent contour lines on a topographic map.

Contour Lines. The lines connecting points of equal elevation on a topographic map.

Contour Map. See Topographic Map.

Contract. An agreement or covenant between two or more persons in which each is bound to do or refrain from doing some act, and each acquires a right to what the other promises.

Contract Award. An owner's notice to a contractor that a bid has been accepted and a contract will be entered into.

Contract Bonds. Performance Bond and Payment Bond, required by some construction contracts.

Contract Documents. The documents that comprise a construction contract, consisting of the agreement between owner and contractor, the conditions of the contract (general, supplementary, and other), drawings, specifications, and addenda issued before the contract is signed; also included are modifications issued after execution of the contract.

Contraction. 1. Becoming smaller from any cause. 2.A shrinking in size when something gets colder. 3. Reduction in the size of a business.

Contraction Joint. Formed, sawed, or tooled groove in a structure to create a weakened plane and regulate the location of cracking resulting from the dimensional change of different parts of the structure.

Contract of Adhesion. A contract offered to a person of inferior bargaining position who has no effective means of negotiating more favorable terms.

Contract of Sale. See Land Contract.

Contractor. (1) An electrical relay to provide power to electrical equipment; an electrically operated switch, usually by a coil or solenoid, containing one or more sets of contacts which control one or more circuits. (2) One who enters into a contract for construction of a building or part of a building; a builder.

Contractor CM. A construction manager who acts as a contractor and who may guarantee the construction cost.

Contractor Manager. The role of the contractor as a business manager.

Contractor's Level. Builder's level.

Contract Sum. The amount stated in the contract for the owner to pay to the contractor for doing the work of the contract.

Contract Time. The amount of time stated in the contract for substantial completion of the work of the contract.

Control Circuit. In a control apparatus or system, the circuit which carries the electric signals directing the performance of the controller, but does not carry the main power.

Control, Compressor. See Motor Control.

Control, Defrosting. See Defrosting Control.

Control Joint. An intentional linear discontinuity in a structure or component, designed to form a plane of weakness where cracking or movement can occur in response to various forces so as to minimize or eliminate cracking elsewhere in the structure; see Control Joint and Relief Joint.

Controller. A device, or group of devices, which serves to govern in some predetermined manner the electric power delivered to the apparatus to which it is conducted.

Control, Low Pressure. Low pressure Control.

Control, Motor. See Motor Control.

Control Panel. A panel, cabinet, or enclosure containing two or more controllers, contractors, relays, or other control devices for the control of electrical circuits, equipment, apparatus or system.

Control, Pressure Motor. High or low-pressure control connected into the electrical circuit and used to start and stop motor. It is activated by demand for refrigeration or for safety.

Control, Refrigerant. See Refrigerant Control.

Controls. 1. A mechanism used to regulate or guide the operation of a machine, apparatus, or system. 2. A functional area of management established to inform the manager when the actual experience of the business is different from that which was planned and to help maintain original budget and plan.

Control System. All of the components required for the automatic control of a process variable.

Control, Temperature. See Temperature Control.

Control Valve. A mechanical device used to regulate an operation or function; a valve which regulates the flow or pressure of a medium which affects a controlled process; such valves are operated by remote signals from independent devices using any of a number of control media such as pneumatic, electric, or electrohydraulic.

Conurbation. The growing together of previously separate towns creating a large community.

Convection. Heat transferred by fluid motion.

Convection Oven. An oven operating by the principal of heat transfer through automatic heated air circulation.

Convection, Forced. See Forced Convection.

Convection, Natural. See Natural Convection.

Convector. A heat exchange device that uses the heat in steam, hot water, or an electric resistance element to warm the air in a room; also called a radiator.

Conventional Installation. The method of installing ceramic tile with portland cement mortar.

Conventional Light-Frame Construction. A type of construction whose primary structural elements are formed by a system of repetitive wood-framing members.

Conventional Loan. A mortgage not insured by the FHA or guaranteed by the VA.

Converter. That which causes change to different state; catalyst; curing agent; promoter.

Convex. A protruding rounded surface, curved or rounded like the exterior of a sphere or circle.

Convex Fillet Weld. A fillet weld having a convex face.

Convexity. The maximum distance from the face of a convex fillet weld perpendicular to a line joining the toes.

Conveyance. The transfer of property from one owner to another, usually by deed.

Conveying System. A device used to move material from one place to another, including elevators, escalators, moving walks, dumbwaiters, and conveyors.

Cooking Unit, Counter-Mounted. An assembly of one or more domestic surface heating elements for cooking purposes designed for flush mounting in, or supported by, a counter, and which assembly is complete with inherent or separately mountable, controls and internal wiring; see Oven, Wall-Mounted.

Cool Colors. Hues or colors in which blue-green predominates; so termed because of the association with ice, water, and sky; also called Cold Colors.

Cooled Loop Run-Around. A method of heat recovery from exhaust air which uses finned-tube coils installed in the incoming and exhaust air supply.

Cooler. Heat exchanger which removes heat from a substance.

Cooler, Bottle. See Bottle Cooler.

Cooler, Nail. A nail with special size and head configuration for use in gypsum board applications.

Cooler, Water. See Water Cooler.

Cooling Coil. A heat exchange between a cooling element and the air stream; the cooling medium may be chilled water or refrigerant gas.

Cooling Tower. A structure in which warm water is circulated for cooling by evaporation by exposure to the air; the water is cooled to the wet bulb temperature of the air.

Cooper. A barrel maker.

Cooperage. 1. Barrels and kegs. 2. A place where barrels are made.

Cooperation. 1. The act of working together for the common good. 2. A behavioral aspect of management; this is achieved by leadership and should be encouraged by management.

Coordinates. A system of magnitudes to fix the position of a point, line, or plane.

COP. Coefficient of Performance.

Copals. Group of resinous substances exuding from various tropical trees; collected from living trees and also dug from the ground as a fossil; includes resins such as amber, congo, kauri, manila, Pontianak, West India gum, and zanzibar.

Cope. To cut or shape the end of a molded wood member so it will cover and fit the contour of an adjoining piece of molding; see Scribe and Scribing.

Coping, Clay. See Clay Coping.

Coped Joint. See Scribing, 2.

Coping. 1. The material or units used to form a cap or finish on top of a wall, pier, pilaster, or chimney; a protective cap at the top of a masonry wall; it should be waterproof and weather resistant and sloped to shed water. 2. Perimeter edging around a swimming pool.

Coping, Limestone. The use of limestone to form a cap or finish on top of a wall, pier, pilaster, or chimney.

Coping, Precast. A precast concrete member used to form a cap or finish on top of a wall, pier, pilaster, or chimney.

Coping Saw. A handsaw with a very narrow blade held under tension in a U-shaped frame; used for cutting curves in wood.

Copolymer. A product of copolymerization; substance obtained when two or more types of monomers polymerize.

Copolymerization. Simultaneous polymerization of two compounds which have properties different from polymer obtained with either monomer separately; see Polymerization.

Copper. A common reddish metallic element that is ductile and malleable and one of the best conductors of heat and electricity.

Copperas. See Ferrous Sulphate.

Copper Bellows. A flexible joint in copper piping that can expand or contract to allow for thermal fluctuations.

Copper Braid. Three or more strands of copper intertwined.

Copper Piping. Pipe and tubing manufactured of copper, classified as Type K, L or M. Type K being the thickest walled, Type M, the thinnest walled.

Copper Plating. Abnormal condition developing in some units in which copper is electrolytically deposited on some compressor surfaces.

Copper Sulfate Test. In test for Mill Scale, copper color indicates absence of mill scale when steel is swabbed with 5 to 10 percent solution.

Copper Sulfate. 1. A chemical compound, $CuSO_4$, used in fungicides, electroplating solutions, textile dyeing, and as a timber preservative 2. An algaecide declining in popularity because of its toxicity and incompatibility with some other compounds found in swimming pools.

Copyright. The exclusive legal right to publish or sell artistic compositions.

Corbel. 1.A masonry unit such as brick or stone which projects beyond the unit below; a spanning device in which masonry units in successive courses are cantilevered slightly over one another; a projecting bracket of masonry or concrete. 2. A projection from the face of a beam, girder, column, or wall used as a beam seat or a decoration.

Corbel Out. To construct a corbel of brick or stone.

Cord. 1. A small, very flexible insulated electrical cable. 2. A flexible thick string or thin rope consisting of woven thin strands. 3. A volume measure of cut firewood of 128 cubic feet (3.63 cubic meters).

Cord, Electric. See Electric Cord.

Cordierite Porcelain. A vitreous ceramic whiteware for technical application in which cordierite, 2MgO $2A_{12}O3$ - $5SiO_2$ is the essential crystalline phase.

Cord Pendant. A cord hanging freely in the air in a vertical position, which has a fixed connection to a permanent wiring enclosure at the upper end and having a suitable cord connection body or lampholder attached to the lower end.

Cordless Tool. A handtool, such as a drill or screwdriver, containing a rechargeable battery.

Cordwood. Wood cut in 4 foot or shorter lengths to be used as fuel; see Cord.

Core. 1. In plywood, the center of the panel; it may be either veneer or lumber. 2. A cylindrical piece of an underground formation cut and raised by a rotary drill with a hollow bit. 3. A hollow space within a concrete masonry unit formed by the face shells and webs. 4. The holes in clay masonry units. 5. The gypsum structure between the face and back papers of gypsum board.

Core, Air. See Air Core.

Core Barrel. A hollow cylinder containing a socket and choker springs for holding a section of drilled rock.

Coreboard. A gypsum board product used primarily in shaftwall systems, normally 1 inch thick or less, either laminated or homogenous, usually manufactured in 24 widths and lengths as per job requirements; may have either square, rounded, or tongue and groove edges.

Core Bracing. Vertical elements of a lateral bracing system such as the walls for stairs, elevators, or duct shafts.

Core Cock. A type of valve through which the flow of water is controlled by a circular core or plug that fits closely in a machined seat; the core has a part bored through it to serve as a water passageway; also called Plug Valve.

Cored Block. See Gypsum Block.

Cored Brick. A brick in which the holes consist of less than 25% of the section.

Core Drill. A rotary drill, usually a diamond drill, equipped with a hollow bit and a core lifter.

Core Drilling. The process of drilling which extracts a cylindrical sample of concrete, rock or soil; sometimes used to install pipe or conduit through an existing concrete or masonry wall.

Core Hardness. The resistance of the core to penetration by a steel punch as measured by ASTM C 473 hardness test.

Core, Magenetic. See Magnetic Core.

Cored Plug. A wooden plug inserted into a drilled hole where a nail, screw or other fastener has been driven.

Core Separation. A split in the gypsumboard core often accompanying an over calcined condition.

Cored Slab. A concrete slab that has been drilled with holes for the installation of plumbing pipe or conduit.

Core Stock. A solid or discontinuous center ply used in panel-type glued structures such as solid or hollowcore doors.

Cored Tile. See Gypsum Tile.

Coring Concrete. To drill concrete to obtain samples for testing or to drill a hole in concrete masonry for conduits or pipe.

Corinthian. The order of Greek architecture characterized by ornate decoration and flared capitals with acanthus leaves.

Cork Running Track. Compressed cork particles used as the base in a running track.

Cork Tile. A flooring material made of a thin sheet of cork cut into a tile shape and attached by adhesive.

Cork Wall Covering. Cork particles that are bound and pressed into sheets and used to deaden sound and add insulation value.

Corner Bead. 1. A metal or plastic strip used to form a neat, durable edge at an outside corner of two walls of plaster or gypsum board. 2. A small, usually curved, wood mold for covering an inside corner.

Corner Block. A large triangular piece of wood or metal used for added strength at the corners of frames or where legs and rails join, see Glue Block.

Corner Boards. Boards used as trim for the corners of a house or other frame structure and against which the ends of the siding are finished.

Corner Bullnose. A type of ceramic tile bullnose trim with a convex radius on two adjacent edges.

Corner Cabinet. A cabinet wall unit that extends down two walls from the inside corner point.

Corner Guard. Type of molding that is mounted on outside corners in a room or space for finishing and for the protection of the corner from damage.

Cornerite. See Corner Reinforcement.

Corner Joint. A joint between two members located approximately at right angles to each other in the form of an L or 90 degrees.

Corner Lath. See Corner Reinforcement.

Corner Post. 1. A vertical post located in the corner of a timber structure. 2. A glazing mullion in the corner of a structure that retains glazing in both walls.

Corner Protection. The act or process of attaching molding to the outside corner of two walls for protection from bumping.

Corner Reinforcement. Plaster reinforcement used at re-entrant or internal angles to provide continuity between two intersecting plaster planes, usually a strip of diamond mesh metal lath bent to form a right angle; also called Cornerite or Corner Lath.

Cornerstone. 1. A stone in a corner of a wall. 2. A ceremonial stone with names of the sponsors, designers, and constructors and the date of construction.

Corner Studs. The arrangement of studs on a corner of a wood frame building that provides nail backing for the lathing or finishes both inside and out.

Corner Tool. In gypsum wallboard finishing, an angular finishing knife to allow the simultaneous application of joint treatment to both sides of a 90 degree interior angle.

Corner Weld. Weld in a joint between two members located approximately at right angles to each.

Cornice. 1. The exterior detail at the meeting of a wall and a roof overhang. 2. A decorative molding at the intersection of a wall and a ceiling. 3. The molded projection which finishes the top of the wall of a building.

Cornice Molding. Molding that is used to form a projection at the top of a wall; a crowning member at the top course of a wall.

Corporation. A group of people authorized by law to act as an individual; an artificial being, created by the State through a charter to engage in a particular kind of business.

Corridor. A passage or hallway from which doors lead into rooms.

Corrosion. Chemical reaction that causes deterioration of metal.

Corrosion-Resistant. Ability of a metal to withstand the effects of corrosion; corrosion resistant metals include any nonferrous metal or any metal having an unbroken surfacing; also includes nonferrous metal or steel with not less than 10 percent chromium or with not less than twenty-hundredths percent copper.

Corrosive Flux. A flux with a residue that chemically attacks the base metal; it may be composed of inorganic salts and acids, organic salts, and acids of activated rosins or resins.

Corrugated Fastener. A type of nail used to hold butt joints together; the slight taper of the corrugations tend to pull the joint together.

Corrugated Glass. Glass that is in a corrugated shape and is translucent but not transparent.

Corrugated Panel. Metal or fiberglass sheeting formed into alternating ridges and valleys in parallel and applied as siding on a building or structure.

Corrugated Roof. Metal or fiberglass formed into alternating ridges and valleys in parallel and mounted on rafters to serve as sheet roofing.

Corrugations. In asphalt paving, a type of pavement distortionion; a form of plastic movement typified by ripples across the pavement surface; these distortions usually occur at points where traffic starts and stops, on hills where vehicles brake on the downgrade, on sharp curves, or where vehicles hit a bump and bounce up and down; they occur in asphalt layers that lack stability, usually caused by a mixture that is too rich in asphalt, has too high a proportion of fine aggregate, has coarse or fine aggregate that is too round or too smooth, or has asphalt cement that is too soft; it may also be due to excessive moisture, contamination due to oil spillage, or lack of aeration when placing mixes using liquid asphalts; see Shoving.

Corundum. Extremely hard crystallized alumina, used as an abrasive.

Cost. The amount paid or charged for something; price.

Cost Approach. A real estate appraisal method by which the replacement cost is estimated and then reduced by the accumulated depreciation.

Cost Estimate. A preliminary statement of approximate cost, determined by one of the following methods. 1. Area and volume method; cost per square foot or cubic foot of the building. 2. Unit cost method; cost of one unit multiplied by the number of units in the project; for example, in a hospital, the cost of one patient unit multiplied by the number of patient units in the project. 3. In-place unit method; cost in-place of a unit, such as doors, cubic yards of concrete, and squares of roofing.

Cost Plus Contract. A type of construction contract where the contract price is the sum of the costs of labor, materials, and subcontracts plus a fixed or percentage fee.

Cost Plus Fee Agreement. A construction contract where the contractor is paid stipulated direct and indirect costs plus a fee. Also called cost plus.

Cottage. A small, simple house, especially in the country.

Cotter Pin. A longitudinally split pin which is inserted into a pre-drilled hole at the end of a rod-type fastener; the two ends of the cotter pin are then spread apart to resist removal of the pin.

Coulomb. An SI unit of electric charge, equal to the quantity of electricity conveyed in one second by a current of one ampere.

Coumarone-Indene Resins. Resins derived as by-products in making coke from coal.

Count. A number identifying yarn size or weight per unit of length, or length per unit of weight, depending on the spinning system used, such as denier, woolen, worsted, cotton, or jute systems.

Counter. A table or case over which transactions are made or work is done.

Counterbalance. A weight or force that balances another

Counterboring. Enlarging a hole so that the head of a screw or bolt inserted in it can be completely covered.

Counterbrace. Bracing installed in opposite directions.

Counterflashing. An inverted L-shaped metal strip built into a wall to overlap base flashing and make a roof or wall watertight; also called Cap Flashing.

Counterflow. Flow in opposite direction.

Counter EMF. Tendency for reverse electrical flow as magnetic field changes in an induction coil.

Counterfort. A buttress or portion projecting from a wall and upward from the foundation to provide additional resistance to thrusts.

Counter Griddle. A cooking device that has been installed in a kitchen counter.

Counter, Laboratory. See Laboratory Counter.

Counter Lavatory. A sink that is installed in a counter top of a base cabinet.

Countersink. A funnel-shaped enlargement of a drilled hole to allow a screw head to be flush with the surface of the drilled material.

Countersunk Plug. A wooden peg used to fill a drilled hole in a wooden surface.

Countertop. The work surface placed on base cabinets in a kitchen, lavatory, or laboratory.

Counter Top Range. A burner assembly mounted on the top of a kitchen counter allowing for additional cabinet space below.

Couple. Where a pair for forces of equal magnitude acting in parallel but opposite directions are capable of causing rotation.

Coupling. A pipe fitting containing female threads on both ends; used to join two or more lengths of pipe in a straight run or to join a pipe and fixture.

Coupling, Compression. A connecting device which when attaching uses a force that pushes together and squeezes.

Coupling Set Screw. A screw used to secure a fitting in place.

Coupling, Split. A coupling that is split longitudinally and is assembled and secured with screws.

Coupling, Threaded. A fitting for joining two lengths of pipe that is threaded for connection to another fitting with connecting threads.

Coupling, Threadless. A fitting for joining two lengths of pipe that is slid over for connection by soldering, welding or cementing.

Course. 1. In masonry, a continuous layer of bricks or block. 2. A continuous row of shingles in a roof. 3. Any layer in a waterproofing system.

Coursed. In masonry, laid in courses with straight bed joints.

Course Textured Wood. Wood with large pores, such as oak, ash, chestnut, and walnut.

Courtyard. A court or enclosure adjacent to a building.

Cove. A trim tile unit having one edge with a concave radius; used to form a junction between the bottom wall course and the floor or to form an inside corner.

Cove Base. A flexible strip of plastic or synthetic rubber used to finish the junction between floor and wall.

Cove Base, Glazed. Cove base tile that has a ceramic coating that is hard, thin, and glossy.

Cove Base, Sanitary. A trim tile having a concave radius on one edge and a convex radius with a flat landing on the opposite edge; used as the only course of tile above the floor tile.

Cove Base, Tile. Tile that is placed in the bottom course of a tiled wall.

Coved Stair Riser. A concave-shaped surface on the vertical part of a stairway riser where the riser meets the horizontal or stair tread.

Cove Molding. Molding that is concave-shaped; used to cover interior angles, such as that between the ceiling and a wall.

Covenant. A promise.

Covenants, Conditions, and Restrictions (CC&Rs). The basic rules establishing the rights and obligations of owners of real property within a subdivision in relation to other owners within the same tract and in relation to an association of owners organized for the operation and maintenance of property owned in common by the individual owners.

Cover. Concrete cover.

Coverage. 1. The scope of an insurance policy. 2. Amount of surface a given quantity of paint will cover. 2. The ability of paint to conceal the surface being painted. 3. The surface area to be continuously coated by a specific roofing material after allowance is made for a specified lap of the material.

Coverage. A measure of the amount of material required to cover a given surface.

Cover, Catch Basin. A cast iron plate or grate on top of a receptacle or reservoir that catches water runoff or aids in drainage.

Covered Electrode. In arc welding, a filler-metal electrode consisting of a metal core wire with a relatively thick covering which provides protection for the molten metal from the atmosphere and improves the properties of the weld metal and stabilizes the arc.

Covering Power. 1. Of paint, the ability to cover the surface to which it is being applied. 2. Of ceramic tile, the ability of a glaze to uniformly and completely cover the fired surface.

Covering Up. Burying, sealing, or otherwise covering work before it has been inspected.

Cover, Manhole. A heavy, usually round, steel or iron cover used to gain access to underground work through a manhole.

Coverplate. A sheet of glass or transparent plastic that sits above the absorber in a flatplate solar collector.

Cover, Pool. A tarp used to cover a pool when not in use.

CP. Chemically Pure.

CPA. Certified Public Accountant.

CPE. Chlorinated polyethylene, a type of single-ply roofing material.

CPM. Critical Path Method.

CPSC. Consumer Product Safety Commission.

CPSE. A type of single-ply roofing material, a self-curing non-vulcanized elastomer. Available as a liquid coating or a membrane sheet. May be reinforced with polyester scrim or laminated to felt backing.

CPVC. Chlorinated Polyvinyl Chloride.

Cr. Credit.

Crab. A hand device used for stretching carpet in a small area where a power stretcher or knee kicker cannot be used.

Crack. 1. To break, split, snap apart or develop fissures. 2. A fracture in the monolithic surface of gypsum board. 3. A separation or fracture occurring in a roof membrane or in a roof deck, generally caused by thermally induced stress or substrate movement; also called a Split. 4. A break in the surface of an asphalt pavement; see Alligatoring, 3.

Crackage. Joint in a structure which permits movement of a gas or vapor through it, even under a small pressure difference.

Crack Control Reinforcement. Reinforcement in concrete construction designed to prevent cracks, often effective in limiting them to uniformly distributed small cracks.

Cracking. Form of paint failure in which breaks in film extend through all coats down to building material.

Cracking a Valve. Opening a valve a small amount.

Crackle Finish. A finish in which alligatoring is produced, allowing the undercoat to show through the cracks. Cracking is produced by rapid drying of topcoat over slow drying undercoat.

Crack Length Method. A method of calculating the quantity of infiltration air into a building; this method requires specific information about dimensions and construction details of windows, doors, and other openings.

Crack Monitor. A two piece adhesive paper pattern that may be attached to a building wall over a crack to record differential structural movement, over time, of the wall on each side of the crack.

Crack Patch. To repair a crack in a surface by using plaster, concrete, asphalt, etc.

Crack Repair. To patch a crack in a surface by using such materials as plaster, concrete, or asphalt.

Crack Sealing. To close the opening between two materials to prevent moisture or air from passing through.

Craft. An occupation or trade requiring artistic skill and manual dexterity.

Craftsman. A person skilled in a trade or handicraft; an artisan; sometimes called Craftsperson.

Craftsmanship. Quality of work done in a trade or handicraft.

Craftsperson. A Craftsman, regardless of gender.

Craft Union. A labor union whose membership is limited to workers in the same trade or craft.

Crane Jib. The projecting arm of a crane.

Crane. A machine for raising, shifting and lowering heavy weights by means of a projecting swinging arm or with the hoisting apparatus supported on an overhead track.

Crane Mobilization. To assemble and put a crane into movement to work at a construction site.

Crane, Travelling. See Travelling Crane.

Crankshaft Seal. Leakproof joint between crankshaft and compressor body.

Crank Throw. Distance between center line of main bearing journal and center line of the crankpin or eccentric.

Crater. A depression at the termination of an arc weld bead.

Cratering. Formation of holes or deep depressions in paint film.

Crawler Crane. A crane mounted on a pair of tracks used to support and propel the crane.

Crawler. Any vehicle propelled by caterpillar tracks, such as a crane.

Crawling. 1. A parting and contraction of the glaze on the surface of a ceramic surface during drying or firing, resulting in unglazed areas bordered by coalesced glaze. 2. Varnish defect in which poor adhesion of varnish to surface in some spots causes it to crawl or gather up into globules instead of covering the surface.

Crawlspace. Space under a floor or in an attic that is accessible by crawling.

Crawl Space Insulation. Insulation that is applied between the first floor joists just above the ground.

Craze Cracks. Fine, random fissures or cracks which may appear in a plaster surface caused by plaster shrinkage; also called check cracking; these cracks are generally associated with a lime finish coat that has not been properly gauged or troweled.

Crazing. 1. Minute, interlacing cracks on the surface of a finish. 2. Hairline cracks in paint film. 3. The cracking which occurs in fired glazes or other ceramic coatings due to critical tensile stress.

Cream Paper. A highly sized and calendered paper used as the face paper in the manufacture of gypsum board.

Credit. 1. A person's financial standing. 2. The sum of money at a person's disposal in a bank or other financial institution. 3. The power to obtain goods or services before payment, based on the trust that payment will be made. 4. A reputation for solvency and honesty in business. 5. The acknowledgement of being paid by an entry on the credit side of an account; the sum entered. 6. An acknowledgement of a contributor's services to the design or construction of a building project.

Creditor. Persons or companies to whom money is owed.

Creditor's Equity. The claims of outsiders against the business; the amounts that the business owes to persons or companies other than the owners.

Creek. A small river.

Creep. 1. A permanent, inelastic, deformation in a material due to changes in the material caused by the prolonged application of structural stress; plastic deformation that proceeds with time when certain materials, such as concrete, are subjected to constant, long-duration stress. 2. The shrinking or stretching of a roofing membrane due to heat or moisture changes.

Creep, Seismic. See Seismic Creep.

Crematorium. A building containing a furnace for cremating the dead; Crematory.

Crematory. See Crematorium.

Creosote. A brownish oily liquid obtained by distillation of coal tar, used as a wood preservative.

Creosote Oil. Distillate, heavier than water, from coal tar, used largely as a wood preservative.

Creosoted Pole. A wooden pole that has been impregnated with creosote to help with its preservation.

Creosote Stain. Creosote, made mostly from wood and coal tars, mixed with linseed oil and drier and thinned with benzine or kerosene.

Crescent. 1. The curved sickle shape of the waning moon, with a convex and a concave edge. 2. Anything of this shape.

Crescent Wrench. An adjustable wrench.

Crest. 1. Of a weir, the surface over which the liquid flows. 2. The ridge of a roof.

Crew Trailer. A trailer provided on a job site for use by workers.

Cricket. A chimney flashing on the uphill side, resembling a small roof ridge, to divert the rainwater around the chimney.

Crimping. Method of texturizing staple and continuous filament yarn to produce irregular alignment of fibers and increase bulk and covering power; also facilitates interlocking of fibers, which is necessary for spinning staple fibers into yarn.

Crimping Pool. A hand operated tool to apply metal corner beads or fasten steel studs to track by clinching part of metal.

Cripple Rafter. The fillers in a roof framing system which connect the valley and hip rafters.

Cripple Stud. 1. A short wood stud occurring in less than full height walls. 2. A short wood stud occurring over door or window headers or under window dsills.

Crisper. Drawer or compartment in refrigerator designed to provide high humidity along with low temperature to keep vegetables, especially leafy vegetables, cold and crisp.

Criterion. (Pl. criteria) A standard on which a judgment or decision may be based.

Critical Path Method (CPM). A construction scheduling system.

Critical Pressure. Compressed condition of refrigerant which gives liquid and gas the same properties.

Critical Temperature. Temperature at which vapor and liquid have same properties.

Critical Vibration. Vibration which is noticeable and harmful to structure.

Crook. The distortion of lumber in which there is a deviation, in a direction perpendicular to the edge, from a straight line from end-to-end of the piece.

Crooked Edges. A curvature of the sides, either convex or concave, measured along the sides of flat surface. The degree of crook is the departure from the straight line between two corners, expressed in percentage of length.

Cross. A pipe fitting with four female threaded openings at right angles to one another.

Crossarm. A horizontal member attached to a utility pole at a 90° angle, used to support cables.

Crossarm Pole. A vertical pole which supports a railroad warning cross-arm on a pivot, which descends when an oncoming train trips an electrical signal to stop oncoming traffic on either side of the railroad tracks.

Crossband. 1. To place the grain of layers of wood at right angles in order to minimize shrinking and swelling. 2. In plywood of three or more plies, a layer of veneer whose grain direction is at right angles to that of the face plies.

Cross Brace. Bracing with two intersecting diagonals; slender diagonal member within a framed wall or partition, to support the wall or partition and to withstand structural loads imposed by wind and suction loads, building loads, movement, and deflection of structure.

Cross Break. A crack or separation of wood cells across the grain of a board; may be caused by unequal shrinkage or by external forces.

Cross Connection. 1. The improper interconnection of potable and contaminated water piping. 2. The improper interconnection of electrical or communication wiring.

Cross-Cut Saw. A hand saw whose teeth are designed for cutting across the wood grain.

Cross Examination. Questions framed to undermine the testimony of a witness.

Cross Fitting. A plumbing fitting consisting of two short pipes meeting at right angles on a pipe run.

Cross Furring. Term used to denote furring members attached to other structural components to support lath in suspended ceilings; generally 3/4 inch steel channels or pencil rods.

Cross Grain. Wood incorporated into a structure in such a way that its direction of grain is perpendicular to the principal loads on the structure.

Cross-Grained Wood. Wood in which the fibers deviate from a line parallel to the sides of the piece; cross grain may be either diagonal or spiral grain or a combination of the two.

Cross Hair. The fine vertical and horizontal lines used for sighting of an optical instrument such as a transit or builder's level.

Cross Joint. See Butt Joint, 4.

Crosslinking. A particular method by which chemicals unite to form films.

Crosslot Bracing. Horizontal compression members running from one side of an excavation to the other, used to support sheeting.

Cross Peen Hammer. A hammer where the face opposite the flat face is shaped for cutting or grooving metal; the peen is flat and parallel to the handle.

Cross-Seam. Seam made by joining the ends of carpet together.

Cross-Section. A drawing representing an orthogonal slice through an object.

Cross-Sectional Area. The area of a section of a member taken at right angles to its axis, such as a steel angle or beam, expressed in square units as square inches or square millimeters.

Cross Slope. A slope that is perpendicular to the direction of travel.

Cross-Spray. Spraying first in one direction and second at right angles.

Crowbar. An iron or steel bar with a flattened wedge at one end, used for prying.

Crowding the Line. The laying of the brick so that the line is not free of the brick wall; this will cause the line to be inaccurate and the wall may bulge or overhang.

Crow Hop. Tile joints that are out of alignment.

Crown. 1. In gypsum wallboard installation, the buildup of joint compound over a joint to conceal the tape; the higher the crown the wider the compound must be feathered to make the joint less visible; also called Crowned Joint or Hump Joint. 2. The increased elevation of the center of a roadway to facilitate drainage to the edges. 3. The highest point of an arch or vault.

Crowned Joint. See Crown, 1.

Crown Glass. Glass sheet formed by spinning an opened globe of heated glass.

Crown Molding. A decorative type of molding used to make ceiling to wall transitions.

Crown Venting. A vent connection no longer permitted on modern installations, consisting of a vent pipe connected at the top or crown of the curve which forms the trap.

Crown Weir. The lower part of the outlet of a plumbing fixture trap.

CRSI. Concrete Reinforcing Steel Institute.

CRT. Cathode Ray Tube.

Crucible. A heat-resistant vessel for melting or calcining materials that require high temperatures.

Cruciform. Forming or arranged in a cross; said of the floor plan of a church with nave and transepts.

Crumbs. Ragged chunks of gypsum on cut outs or cut ends of gypsum wallboard.

Crushed Gravel. The product resulting from artificial crushing of gravel with substantially all fragments having at least one fracture face.

Crushed Marble. Marble that has been crushed to smaller and more uniform sizes so substantially one face of each stone is fractured.

Crushed Stone. The product resulting from the artificial crushing of rocks, boulders, or large cobblestones substantially all faces of which have resulted from the crushing operation.

Crusher Run. Gravel, rock, boulders, or blasted rock that has been reduced in size by a machine, but has not been sorted for size; the total unscreened product of a stone crusher.

Crust. The outer part of the earth, the lithosphere, the outer 80 kilometers of earth's surface made up of rocks, sediment, and basalt; general composition is silicon-aluminum-iron.

Cryogenics. 1.The branch of physics dealing with the production and effects of extremely low temperatures. 2. Refrigeration which deals with producing temperatures of minus 250° F. and lower.

Cryogenic Fluid. Substance which exists as a liquid or gas at ultra-low temperatures, minus 250° F. or lower.

Crypt. A vault under the main floor of a church.

Cryptometer. An instrument used to measure the opacity of paint.

Crystalline Glaze. 1. Glazed tile with an extra heavy glaze produced for use on counter tops and light duty floor surfaces where abrasion or impacts is not excessive. 2. A glaze that contains microscopic crystals.

Crystallizing Lacquer. Novelty finish which crystallizes forming unusual crystal and floral patterns as it dries.

CS. Commercial Standard.

CSTC. Ceiling Sound Transmission Class.

C Stud. A roll-formed metal channel stud.

C. Switch (Isolating). A switch intended for isolating an electric circuit from the source of power. It has no interrupting rating and is intended to be operated only after the circuit has been opened by some other means.

CTI. Ceramic Tile Institute.

C to C. Center to center.

Cubage. The cubic contents of a vessel, space, or building, expressed in cubic units, such as cubic feet or cubic meters; also called Cubature.

Cubature. See Cubage.

Cubicle Adapter. A device for connecting partitions into different positions.

Cubicle, Hospital. See Hospital Cubicle.

Cubicle, Office. See Office Cubicle.

Cul De Sac. A dead end road no more than 400 feet long with a turn-around of no less than 80 feet in diameter.

Culling. Sorting of the brick for quality, size and color.

Culls. The rejected bricks in culling.

Cultured Marble. A plastic imitation marble for interior use.

Culvert. A drain pipe or small bridge for drainage under a road or structure.

Culvert Formwork. The temporary support for the pouring, placing and curing of a concrete drain pipe constructed under a road or embankment to provide a waterway.

Culvert Reinforcement. The placing of metal or steel bars in concrete forms for drainage systems or pipe; usually refers to a box culvert shape constructed under a road or embankment to provide a waterway.

Cumulative Errors. Errors that are always positive or always negative, as contrasted with compensating errors.

Cup. A distortion of a board in which there is a deviation flatwise from a straight line across the width of the board.

Cupola. A small domed structure adorning a roof.

Cupped Taper. In gypsum wallboard, a condition where the outer edge of the taper is in the same plane as the surface causing a ridge to appear in the tapered edge.

Cuprous. Of or like copper.

Curb. 1. A protective rim. 2. A concrete edging raised above a roadway and forming the edge of a drainage gutter.

Curb and Gutter. Concrete or stone structure that forms the edging of a sidewalk, separating it from the paved street; the adjacent, usually integral, gutter provides a drainage channel.

Curb Cock. A valve placed on the water service usually near the curb line; also called the Curb Stop.

Curb Cut. An interruption of a curb at a driveway pedestrian way.

Curb Form. A retainer or mold with the necessary shape to make a concrete curb.

Curb Formwork. The temporary support for a freshly placed or poured curb system.

Curb Granite. Extremely hard rock used in a curb system; available in sections.

Curb Inlet Frame. The steel or metal framing surrounding an opening in a curb through which water flows and drains.

Curb Inlet. The opening in a curb through which water flows and drains.

Curb Ramp. A sloping pedestrian way, intended for pedestrian traffic, which provides access between a walk or sidewalk to a surface located above or below an adjacent curb face.

Curb Removal. The act or process of the demolition of a curb into manageable pieces.

Curb Roof. See Gambrel Roof.

Curb Stop. See Curb Cock.

Curb, Terrazzo. A curb made of decorative mosaic material made by embedding small pieces of marble or granite in mortar and polishing.

Cure. 1. The process of concrete hardening, rubber vulcanizing, or adhesive achieving its maximum strength. 2. To change the properties of an adhesive by chemical reaction, which may be condensation, polymerization, or vulcanization, and thereby develop maximum strength; generally accomplished by the action of heat or a catalyst, with or without pressure. 3. Toughening or hardening of paint film. 4. To provide conditions conducive to completion of the hydration process in portland cement concrete or plaster.

Cured. Completely dry; moisture free.

Curing. Maintenance of humidity and temperature of freshly placed concrete during some definite period following placing, casting, or finishing to assure satisfactory hydration of the cementitious materials and proper hardening of the concrete; the hardening of concrete or plaster.

Curing Agent. Hardener; promoter.

Curing Blanket. A built-up covering of sacks, matting, hessian, straw, waterproof paper, or other suitable material placed over freshly finished concrete to prevent premature dehydration.

Curing Compound. A liquid that can be applied as a sprayed coating to the surface of newly placed concrete to retard the loss of water or, in the case of pigmented compounds, also to reflect heat so as to provide an opportunity for the concrete to develop its properties in a favorable temperature and moisture environment; forms a water-resistant layer to prevent premature dehydration of the concrete.

Curing Concrete. To control the humidity and temperature of freshly finished concrete to assure the proper drying and hardening of the concrete.

Curing, Electrical. See Electrical Curing.

Curing Paper. Waterproof paper placed over freshly finished concrete, to help control the humidity and temperature, aiding in the proper curing of concrete.

Curing Period. The amount of time that concrete should be kept damp after placing, usually about a week.

Curing, Steam. See Steam Curing.

Curing Temperature. The temperature to which an adhesive or an assembly is subjected to cure the adhesive.

Curling. The distortion of an originally essentially linear or planar member into a curved shape such as the warping of a slab due to creep or to differences in temperature or moisture content in the zones adjacent to its opposite faces.

Curly-Grained Wood. Wood in which the fibers are distorted so that they have a curled appearance, as in birdseye wood; the areas showing curly grain may vary up to several inches in diameter.

Current. 1. Electrical flow through conductors, measured in amperes. 2. Belonging to the present time, as current week or month.

Current Assets. Cash and other assets that can be easily and quickly converted to cash, usually within one year.

Current Liabilities. Liabilities to be paid within a certain time, usually one year.

Current-Limiting Overcurrent Protective Device. A device which, when interrupting a specified circuit, will consistently limit the short-circuit current to a specified magnitude substantially less than that obtainable in the same circuit if the device was replaced with a solid conductor having comparable impedance.

Current Ratio. The ratio of current assets to current liabilities.

Current Regulator. An automatic electrical control device for maintaining a constant current in the primary of the welding transformer.

Current Relay. Device which opens or closes a circuit; made to act by a change of current flow in that circuit.

Curtain. 1. A single layer of vertical and horizontal reinforcing bars in a wall. 2. Window drapery.

Curtain Drain. See Intercepting Drain.

Curtain, Gym. See Gym Curtain.

Curtain Rod. A horizontal bar that supports window coverings.

Curtain Wall. An exterior non-bearing wall between columns, sometimes containing windows or are all glass.

Curtilage. The ground or yard surrounding a house up to the boundary fence.

Curvature Factor. A factor applied to the allowable bending stress of glue-laminated beams where a curved portion occurs.

Curvature Friction. The friction resulting from bends or curves in the specified prestressing tendon profile.

Curved Curb Form. A curve shaped support for the placing, pouring and curing of a concrete curb.

Curved Stair Terrazzo. Terrazzo tile used in a curved stair system.

Curved Wall Form. A curved shaped support for the placing, pouring, and curing of concrete in a curved wall system.

Cushion-Edged Tile. Tile on which the facial edges have a distinct curvature that results in a slightly recessed joint.

Cushioning, Carpet. See Carpet Padding.

Cusp. A projecting point between small arcs in Gothic tracery.

Custom and Usage. The way the majority of those in the trades and professions commonly conduct themselves in doing their particular work.

Custom Door. A door that has been manufactured to fit a specific size opening or manufactured from specific materials.

Cut. Dispersion of a certain number of pounds of shellac or resin per gallon of volatile liquid; for example, a 4 pound cut of shellac contains 4 pounds of dry shellac and 1 gallon of alcohol.

Cut and Cover. A work method which involves excavation in the open, and placing of a temporary roof over it to carry traffic during further work.

Cut and Fill. Excavated material removed from one location and used as fill material in another location.

Cutback. Asphalt or tar that has been diluted with solvents and oils so the resulting material becomes fluid.

Cutback Asphalt. Asphalt cement which has been liquefied by blending with petroleum solvents, as for the RC and MC cutback asphalts; upon exposure to atmospheric conditions the solvents evaporate, leaving the asphalt cement to perform its function.

Cut End. The end of the gypsum board with the exposed core.

Cut-In. The temperature value or the pressure value at which the control circuit closes.

Cut in the Sash. Painting the window sash; ordinarily done with a brush, often called a sash tool, which permits the painter to get a clean edge.

Cut Joints. Masonry bed and head joints cut flush with trowel.

Cut Loop Pile. In carpet making, pile surface in which the tufts have been cut to reveal the fiber ends.

Cutoff. In roofing installation, the procedure to prevent lateral water or moisture entering the insulation where it terminated at the end of the work day; usually a felt strip hot-mopped to the deck and to the insulation edge and to the horizontal insulation surface; at the start of the next day's work the strip is taken off.

Cut Out. 1. A mechanical or electrical device used to break an electrical circuit because of an overload. 2. An opening in a wall or surface for access. 3. A piece stamped out of metal. 4. An opening in a countertop for access or installation of a piece of equipment.

Cut-Out. Temperature value or pressure value at which the control circuit opens.

Cut Stock. Softwood stock comparable to dimension stock in hardwoods; see Dimension Stock.

Cut Stone. Building stone cut to size and shape for specific applications and designated locations in a building or structure.

Cutting. 1. Excavating. 2. Sawing. 3. Slicing.

Cutting Attachment. A device which is attached to gas welding torch to convert it into an oxygen-cutting torch.

Cuttings. In hardwoods, a portion of a board or plank having the quality required by a specific grade or for a particular use; obtained from a board by crosscutting or ripping.

Cutting Tip. That part of an oxygen-cutting torch from which the gases issue.

Cutting Torch. A nozzle or device used in oxygen cutting for controlling and directing the gases used for preheating and the oxygen used for cutting the metal.

CW. Cold Water.

Cyanic. Containing blue or pertaining to blue color.

Cyanuric Acid. Acid used in pool water to prevent chlorine loss.

Cybernetics. The science of communications and automatic control systems in both machines and living things.

Cycle. 1. One complete performance of an electrical oscillation or current alternation. 2. A series of events or operations which have the tendency to repeat in the same order.

Cyclone. A storm or system of winds that rotate about a center of low atmospheric pressure.

Cyclopean Aggregate. Concrete aggregate where the individual pieces are over 100 pounds.

Cyclotron. An apparatus in which charged atomic and subatomic particles are accelerated by an alternating electric field while following an outward spiral or circular path in a magnetic field.

Cylinder. 1. The surface traced by a straight line moving parallel to a fixed straight line and intersecting a fixed planar closed curve. 2. Of a lock, the cylindrical-shaped assembly containing the tumbler mechanism and the keyway. 3. A device which converts fluid power into linear mechanical force and motion; this usually consists of movable elements such as a piston and piston rod, plunger or ram, operating within a cylindrical bore. 4. A closed container for fluids. 5. A portable cylindrical container used for transportation and storage of a compressed gas.

Cylinder Glass. Glass sheet produced by blowing a large, elongated glass cylinder, cutting off its ends, slitting it lengthwise, and opening it into a flat rectangle.

Cylindrical Commutator. Commutator with contact surfaces parallel to the rotor shaft.

Cylindrical Lock. A door lock that can be installed in a cylindrical hole rather than in a mortise; also called a Tubular Lock.

Cylinder Piling. Concrete filled steel pipes or tubes, driven through the ground to reach bedrock and used as a foundation for tall buildings and in underpinning.

Cylinder Plug. The round part of a door lock containing the keyway and rotated by the key to transmit motion to the bolt.

Cylinder, Refrigerant. Refrigerant Cylinder.

Cylinder Test. A test to determine the compressive strength of concrete.

Cyma. A type of cove molding in which the surface of the face changes from concave to convex; a Cyma Recta or Cyma Reversa; commonly made of wood, plaster, or masonry.

Cyma Recta. An ogee molding combining a concave curve with a convex curve; when the concave curve is on top and the convex curve is on the bottom it is a cyma recta; see Cyma Reversa.

Cyma Reversa. An ogee molding combining a concave curve with a convex curve; when the convex curve is on top and the concave curve is on the bottom it is a cyma reversa; see Cyma Recta.

D

D&M. Dressed and Matched.

Dado (pl. Dadoes). 1. Part of a column base. 2. The lower part of an interior wall when differently surfaced. 3. A groove in a piece of wood made by dadoing.

Dado Joint. A joint in which one piece is grooved to receive the piece which forms the other part of the joint.

Dairy. A building or room for the storage, processing, or sales of milk and milk products.

Dalton's Law. Vapor pressure created in a container by a mixture of gases is equal to the sum of individual vapor pressures of the gases contained in mixture.

Damages. 1. Injuries. 2. An amount of money awarded to a plaintiff to compensate for loss caused by the wrongful conduct of a defendant.

Dammar. A natural resin used extensively in the preparation of varnishes and lacquers; usually classified according to the place from which it is shipped to market, for example, Singapore dammar or Batavia dammar.

Damp Check. See Damp Course.

Damp Course. A course or layer of impervious material which prevents capillary entrance of moisture from the ground or a lower course; also called Damp Check.

Damper. A flap to control or obstruct the flow of air or other gasses; specifically, a metal control flap in the throat of a fireplace, or in an air duct; controls that vary airflow through an air outlet, inlet, or duct; a damper position may be immovable, manually adjustable, or part of an automated control system.

Damping. 1. Dissipation of structure-borne noise by conversion to some other form of energy, usually heat; usually accomplished by using a material with a high internal energy-absorbing capacity. 2. A rate at which natural vibration decays as a result of absorption of energy.

Damp Location. A location subject to a moderate degree of moisture, such as some basements, some barns, some cold storage warehouses, and the like.

Dampproofing. The treatment of concrete or mortar to help prevent the passage or absorption of water in the absence of hydrostatic pressure; not necessarily waterproof.

Dap. To cut and form a recess in timber to form a dapped joint; also called a let-in joint, as where 1" x 6" diagonal bracing is cut into 2" x 4" studs.

Darby. 1. A hand-manipulated straightedge, usually 3 to 8 feet. (1 to 2.5 m) long, used in the leveling operation of the early stage of concrete placement, preceding supplemental floating and finishing. 2. A stiff straightedge, about 4 inches wide and 42 inches long with handles, of wood or metal, used to level the surface of wet plaster.

Darkroom. A room for photographic work, with normal light excluded.

Darkroom Dryer. A mechanical drying device used by a photographer in a darkroom to dry prints.

Darkroom Equipment. Equipment used for the developing of photographs.

Darkroom Sink. A basin in a photographic darkroom for the dispensing or holding of liquids.

Dash Bond Coat. A thick slurry of portland cement, sand, and water flicked on surfaces with a paddle or brush to provide a base for subsequent portland cement plaster coats; sometimes used as a final finish on plaster.

Dash Brush. A long-bristled brush for flinging a plaster mixture on the wall as a dash coat.

Datum. 1. The fixed starting point for surveying or measuring. 2. A piece of information.

Daub. A glob of adhesive.

Daylight. A painting defect where there is a lack of coverage of a coat of paint.

Daylighting. Illuminating the interior of a building by natural means.

Days. Calendar days are consecutive days on the calendar; working days are calendar days less non-working days such as week ends and holidays.

db. Decibel.

DB. Dry Bulb.

DBA. Doing Business As.

DBIA. Design/Build Institute of America.

Dbl Conduit. Double thick conduit.

Dbl Duct. Double thick metal duct.

DC. Direct Current.

D Cracking. The progressive formation on a concrete surface of a series of fine cracks at rather close intervals, often of random patterns, but in slabs on grade paralleling edges, joints, and cracks and usually curving across slab corners; also called D-Cracks and D-Line cracks.

D Cracks. See D Cracking.

DE. Diatomaceous Earth.

Deadbolt. A lock bolt having no spring action nor bevel, and which is operated by a key or a turn piece.

Dead Burned. Removal of all water content during calcining of gypsum.

Dead End. A branch leading from a soil, waste or vent pipe, building drain sewer which is terminated at a developed length of 2 feet or more by means of a plug or other fitting. A dead end is not used to admit water or air into the piping system.

Deadening. Construction intended to prevent the passage of sound.

Dead Flat. In paint, having no gloss at all.

Dead-Front. As applied to switches, circuit breakers, switchboards, control panels, and panelboards, so designed, constructed, and installed that no current-carrying parts are normally exposed on the front.

Dead Knot. Loose knot in wood, usually dark in color.

Dead Level Asphalt. A roofing asphalt that has a softening point of 140° F (60° C).

Dead Level. An absolutely horizontal roof. slope.

Dead Load. The load due to the vertical weight of all permanent structural and nonstructural elements of a building, such as walls, floors, and roofs.

Deadlock. A lock equipped with a deadbolt only.

Deadman. A large or heavy object buried in the ground as an anchor.

Dead Man's Switch. One that stays in on or off position only as long as it is held there manually.

Dead Space. 1. A space with an abundance of absorptive surfaces that does not support sound. 2. A space between walls or a floor and ceiling.

Deaeration. Act of separating air from substances.

Deal. Softwood timber.

Debris. Accumulated rubbish, trash and fragments of waste.

Debt. Money that is owed.

Debt, Long Term. See Long Term Debt.

Debtors. Persons or companies that owe money to the business; accounts or notes receivable.

Debt, Short Term. See Short Term Debt.

Decal. Decalcomania.

Decalcomania. Pictures,diagrams and designs, on specially prepared paper, that can be transferred to glass or other surfaces; also called a Decal.

Decay. The decomposition of wood or other substance by fungi or bacteria.

Dechlorination. The removal of chlorine from water.

Decibel (db). A standard unit of measure for classifying the transmission of airborne sound levels; the ratio of sound pressure to a base level chosen at the threshold of hearing and ranging up to 130 for the average pain level.

Deciduous. Trees which annually lose their leaves.

Decimal. Any real number expressed in base 10.

Decimal Fraction. A fraction or mixed number in which the denominator is a power of 10, usually expressed by use of the decimal point, as 1/4 = 25/100 = 0.25.

Deck. 1. The roof surface to be covered, the substrate or substrata. 2. A small platform used for walking. 3. The form on which concrete for a slab is placed. 4. A floor or roof slab.

Deck, Coil. See Coil Deck.

Decking. 1. A material used to span across beams or joists to create a floor or roof surface. 2. Heavy plank floor of a pier or bridge. 3. Surfaced area surrounding a swimming pool.

Decking, Cellular. See Cellular Decking.

Decking, Metal. See Metal Decking.

Decking, Open. See Open Decking.

Decking, Wood. See Wood Decking.

Deck Insulation. Thin sheets of insulation placed on a concrete surface before a flooring system is installed.

Deck Level Pool. Water surface level with the deck.

Deck Mud. See Floor Mud.

Deck Paint. An enamel with a high degree of resistance to mechanical wear; designed for use on such surfaces as porch floors.

Declaration of Homestead. See Homestead, 3.

Declining Balance. A form of accelerated depreciation where a larger amount is taken in the beginning years gradually declining to a smaller amount in the later years.

Decor. The furnishing, layout, style, and decoration of building interiors.

Decorated. Adorned, embellished, or made more attractive by means of color or surface detail.

Decoration. 1. The act of decorating. 2. The finish used over the gypsumboard face.

Decoration, Inglaze. See Inglaze Decoration.

Decoration, Overglaze. See Overglaze Decoration.

Decoration, Underglaze. See Underglaze Decoration.

Decorative Painting. Architectural painting; aesthetic painting.

Decorative Tile. Tile with a ceramic decoration on the surface.

Decoupling. Separating elements to retard the transmission of structural-borne sound, thermal conductance, or physical loads.

Dedication. The appropriation of land or easements by a private owner for public purposes or use.

Deductive Alternate. An alternate bid that, if accepted, reduces the contract price.

Deed of Trust. See Trust Deed.

Deed Restrictions. Restrictive covenants often included in a deed which may limit future uses of the land, and other restrictions such as height, size, and aesthetics, as long as they are not against the public interest.

Deeds. A general term which refers to all documents conveying property, from one person to another.

Deep Color. Intense or strong color with no apparent presence of black.

Deep Seal-P Trap. A P-trap with a water seal depth of more than 4 inches.

De Facto. In reality; actually; in fact, whether by right or not.

Default. A material failure to perform the requirements of a contract.

Defect. An imperfection.

Defendant. A party against whom a legal claim is made in a civil law suit.

Deferred Maintenance. Repair and upkeep that has been neglected or postponed.

Deficit. An economic shortfall, where expenses exceed income; a debit balance in the retained earnings account.

Deflection. 1. A variation in position or shape of a structure or structural element due to effects of loads or volume change, usually measured as a linear deviation from an established plane rather than an angular variation. 2. Displacement or bending of a structural member due to application of external force.

Deformation. The change in shape of a body brought about by the application of a force. Deformation is proportional to the force within the elastic limits of the material.

Deformation, Eutectic. See Eutectic Deformation.

Deformed Bar. A steel reinforcing bar manufactured with deformations (bumps, lugs, or ridges) to provide a locking anchorage with the surrounding concrete.

Deformed Reinforcement. Deformed reinforcing bars, bar and rod mats, and deformed wire fabric.

Defrost Device. Refrigerating cycle in which evaporator frost and ice accumulation is melted.

Defrosting. Process of removing frost accumulation from evaporators.

Defrosting Control. Device to automatically defrost evaporator; it may operate by means of a clock, door cycling mechanism, or during off portion of refrigerating cycle.

Defrosting Type Evaporator. Evaporator operating at such temperatures that ice and frost on surface melts during off part of operating cycle.

Defrost Timer. Device connected into electrical circuit which shuts unit off long enough to permit ice and frost accumulation on evaporator to melt.

Degradation. A decline in quality; degeneration.

Degreaser. Chemical solution for grease removal.

Degreasing. Solution or solvent used to remove oil or grease from refrigerator parts.

Degree. 1. A unit on one of the temperature scales. 2. A unit of measurement of angles, equal to 60 minutes; a right angle is 90 degrees.

Degree Day. Unit that represents one degree of difference from inside temperature and the average outdoor temperature for one day and is often used in estimating fuel requirements for a building.

Degree Day Method. A method of computing fuel requirements for HVAC systems, see Degree Day.

Degree of Saturation. The amount of water present in the air relative to the maximum amount it can hold at a given temperature without causing condensation.

Dehumidification. The removal of moisture from air.

Dehumidifier. A machine which removes moisture from the air.

Dehydrated Castor Oil. A drying oil prepared from castor oil.

Dehydrated Oil. Lubricant which has had most of its water content removed, also called Dry Oil.

Dehydration. The process of removing water from a substance.

Dehydrator. See Drier, 1.

Dehydrator Receiver. Small tank which serves as liquid refrigerant reservoir and which also contains a desiccant to remove moisture; used on most automobile air conditioning installations.

Deice Control. Device for operating a refrigerating system in such a way as to provide melting of the accumulated ice and frost.

De Jure. By right or of right.

Delamination. 1. Separation of one layer from another. 2. The separation of layers in laminated wood because of failure of the adhesive, either within the adhesive itself or at the interface between the adhesive and the adherend. 3. The failure of a built-up roofing membrane, resulting in the separation of the felt plies, resulting in cracking and wrinkling. 4. The separation of the paper plies or surface coverings on gypsum board.

Delay. Time lost in the construction schedule by any reason.

Delicatessen Case. A refrigerated cabinet used for the storing and displaying of perishable foods.

Delignification. Removal of part or all of the lignin from wood by chemical treatment.

Deliquescence. The natural absorption of atmospheric moisture by a solid substance.

Delivery Cart. A heavy or lightweight, two or four-wheeled vehicle used to transport goods.

Delivery Tolerances. In asphaltic concrete making, permissible variations from the exact desired proportions of aggregate and bituminous material as delivered into the pugmill.

Delta. 1. The 4th letter in the Greek alphabet, the capital delta being shaped like an equilateral triangle. 2. Anything so shaped. 3. The alluvial deposit at the mouth of a river. 4. A numbered marking on contract drawings to identify changes.

Delta Transformer. Three-phase electrical transformer which has ends of each of three windings electrically connected to form a triangle.

Delustered Nylon. Nylon on which the normally high sheen has been reduced by surface treatment.

Demand Factor. The demand of any system; the ratio of the maximum demand of the system to the total connected load of the system; the loads of a system are practically never on at the same time due to the many uses of the power; somewhere between the maximum connected load and the actual usage is a load that is considered the maximum demand.

Demand Meter. Instrument which measures the kilowatt-hour usage of a circuit or group of circuits.

Demise. Transfer property by will, grant, or lease.

Demised Premises. The property leased.

Demising Wall. See Party Wall, 3.

Demography. The study of the statistics of a human population in respect to trends in size, density, age, and economic characteristics, to identify markets.

Demolish. To raze or tear down a building or structure.

Demountable Partition. An assembly designed to be dismantled and reassembled with a minimal loss of components.

Demur. To take exception; object.

Demurrage. A charge for detaining a ship, freight car, or truck during the period of unloading.

Demurrer. A response in a court proceeding in which the defendant does not dispute the truth of the allegation but claims it is not sufficient grounds to justify legal action; an objection.

Den. A private comfortable room for pursuit of hobbies or recreation.

Denatured Alcohol. Grain or ethyl alcohol made unsuitable for beverage purposes by adding compounds of a poisonous nature.

Dendriform. Resembling a tree in structure.

Dendrochronology. The science of dating events and variations in environment in former periods by comparative study of growth rings in trees and aged wood.

Dendroid. Resembling a tree in form.

Dendrology. The study of trees.

Denier. System of yarn count used for synthetic fibers; the weight in grams per 9,000 meters of yarn length; one denier equals 4,464,528 yards per pound or 279,033 yards per ounce.

Dense-Graded Aggregate. An aggregate that has a particle size distribution such that when it is compacted, the resulting voids between the aggregate particles, expressed as a percentage of the total space occupied by the material, are relatively small.

Densification. The act of increasing the density of a mixture during the compaction process.

Density. 1. The quantity per unit volume, unit area, or unit length; in wood it is expressed as pounds per cubic foot, kilograms per cubic meter, or grams per cubic centimeter at a specified moisture content. 2. Closeness of texture or consistency of particles within a given substance.

Density, Pile Yarn. See Pile Yarn Density.

Dental Equipment. Equipment used by dentists for dental work on patients.

Dental Light. An adjustable lamp used by dentists, having an arm-like extension enabling it to be maneuvered in any direction, and which focuses light in one direction.

Dentil. One of a series of small projecting rectangular blocks forming a molding, part of a cornice.

Deodorizer. Device which absorbs or adsorbs various odors, usually by the principle of absorption; activated charcoal is commonly used.

Deposit. 1. A natural resource such as iron ore. 2. Settled material, such as sediment. 3. A returnable sum paid usually for safekeeping or to guarantee the return of something.

Deposited Metal. Filler metal that has been added during a welding operation.

Deposition. 1. The extra-judicial testimony of a witness given under oath. 2. The written record thereof.

Depository, Night. See Night Depository.

Depreciation. The decrease in value of a property from any cause.

Derrick. A hoisting frame with a long beam, ropes, gear, and pulleys that can lift a heavy load and swing it to the side.

Desiccant. 1. A drying agent. 2. Substance used to collect and hold moisture in refrigerating system; common desiccants are activated alumina and silica gel.

Design. To conceive or plan out in the mind.

Design/Build. A project delivery system in which the same entity is responsible for both design and construction.

Design Concept. An architectural idea that is depicted on the drawings and in the specifications with the expectation that it will be faithfully executed by the contractor.

Design Development Drawings. Drawings prepared by architects for the Design Development Phase.

Design Development Phase. One of the standard phases of architectural service (Schematic Design Phase, Design Development Phase, Construction Documents Phase, Bidding or Negotiation Phase, and Construction Phase-Administration of the Construction Contract).

Design Displacement. The design-basis earthquake lateral displacement, excluding additional displacement due to actual and accidental torsion, required for design of the seismic isolation system.

Design Filter Rate. See Design Rate of Flow.

Design Head. The total head requirement of the circulation system at the design rate of flow.

Design Heat Load. The total heat loss from a building under the most severe winter conditions likely to occur.

Design Pressure. Highest or most severe pressure expected during operation; sometimes used as the calculated operating pressure plus an allowance for safety.

Design Professional. A person educated and skilled in the field of building design or the related structural and environmental systems.

Design Rate of Flow. In a pool, spa, or hot tub, the rate of flow in gallons divided by the number of minutes in the turnover time; also called the Design Filter Rate.

Design Temperature. A temperature close to the lowest expected for a location, used to determine design heat load.

Destructive Distillation. Distilling a product at a temperature so high that products obtained are of different chemical composition than existed in the original material.

Destructive Testing. Testing of building materials in a manner that destroys the material.

Detail. 1. Extended treatment or attention to particular items. 2. Drawing of a small portion of an assembly.

Detectable Warning. A standardized surface or feature built in or applied to walking surfaces or other elements to warn visually impaired persons of hazards in the path of travel.

Detector, Leak. See Leak Detector.

Detector, Smoke. See Smoke Detector.

Detergent. Cleaning agent.

Deterioration. Depreciation; loss of value of real property due to physical wear and tear.

Determinate. 1. Having defined limits; definite. 2. Statically Determinate; see Statically Determinate Structures.

Developed Length. The length as measured along the center line of the pipe and fittings.

Development Length. In reinforced concrete, the length of steel reinforcing required to develop the required bond between the reinforcing and the adjacent concrete.

Device. 1. A piece of equipment or a mechanism designed for a special purpose or function. 2. An ornamental design. 3. A unit of an electrical system which is intended to carry but not utilize electric energy.

Device Cover. A protective or ornamental removable cover over an electrical box or device.

Devil's Float. In plastering, a wooden float with two nails protruding from the toe, used to roughen the surface of the brown coat.

Dew. Condensed atmospheric moisture deposited in small drops on cool surfaces.

Dewatering. To remove water from the ground or excavations with pumps, wellpoints, or drainage systems.

Dew Point. Temperature at which vapor, at 100 percent humidity, begins to condense and deposit as liquid on cold surfaces.

Dew-Point Temperature. The temperature at which condensation occurs.

Dextral. Right handed.

DF. 1. Douglas Fir. 2. Drinking Fountain.

DG. Decomposed Granite.

DHI. Door & Hardware Institute.

Diagnostic Architecture. The science and practice of analyzing building failures to determine the causes and devise repair methods.

Diagonal. 1. Running in an oblique direction from a reference line. 2. Inclined member of a truss or bracing system used for stiffening or wind bracing.

Diagonal Bond. This is a form of raking bond in which the bricks are laid in an oblique direction in the middle section of a thick wall; the bricks may also be laid in this fashion in paving.

Diagonal Bracing. That form of bracing that diagonally connects joints at different levels.

Diagonal-Grained Wood. 1. Wood in which the annual rings are at an angle with the axis of a piece as a result of sawing at an angle with the bark of the tree or log. 2. A form of cross grain.

Diagonal Sheathing. Wood sheathing with the individual boards running at a 45 degree angle to the studs, joists, or rafters.

Diameter. A chord passing through the center of a figure or body, such as a circle.

Diamond Mesh Window Guard. Guard fabricated of diamond-shaped mesh to provide protection over the face of a window to prevent damage to glass and to prevent intrusion.

Diamond Mesh. One of the common types of metal lath having a characteristic geometrical pattern produced by the slitting and expansion of metal sheets.

Diaper. 1. An allover pattern consisting of one or more small repeated units of design, as geometric figures, connecting with one another or growing out of one another with continuously flowing or straight lines. 2. Any repeated and continuous pattern in brickwork, usually applied to diamond or other diagonal patterns.

Diaphragm. 1.Flexible material usually made of thin metal, rubber or plastic. 2. A thin, usually rectangular or square element of a structure that is capable of resisting lateral forces in its plane, such as a floor or roof.

Diaphragm Action. A bracing action that derives from the stiffness of a thin plane of material when it is loaded in a direction parallel to the plane; diaphragms in buildings are typically floor, or roof surfaces of plywood, reinforced masonry, steel decking, or reinforced concrete.

Diaphragm Pump. A water pump used to continuously remove water from excavations containing mud and small stones.

Diatomaceous Earth (DE). Sedimentary rock deposits consisting of or abounding in diatoms or their siliceous remains; diatomite; used in water filters.

Diatomaceous Earth Filter. Pool filter using DE as filtering medium; one designed to filter water through a thin layer of filter aid such as diatomaceous earth or volcanic ash; diatomite filters may be of the Pressure or Vacuum type.

Dichlorodifluoromethane. Popular refrigerant known as Freon 12.

Dichlorodifluromethane. Refrigerant commonly known as R-12.

Dichromatic. Having two colors.

Die. 1. To cease living. 2. To stop. 3. An engraved metal device for stamping coins or medals. 4. A perforated steel block against which a material pressed to form a shape by extrusion. 5. An internally threaded tool used to cut threads on a pipe or rod.

Die Casting. Process of molding low-melting temperature metals in accurately shaped metal molds.

Diecast Set Screw. A set screw cast by forcing molten metal into a mold.

Dielectric Fluid. Fluid with high electrical resistance.

Diesel Generator. An electrical generator operated by diesel fuel.

Different Systems. Those which derive their supply from different sources, or from individual transformers or banks of transformers which do not have their secondary windings interconnected, or from individual service switches.

Differential. The temperature or pressure difference between cut-in and cut-out temperature or pressure of a control.

Differential Settlement. The uneven sinking of different parts or sections of a building's foundations.

Diffuse Radiation. Sunlight that is scattered from air molecules, dust and water vapor and comes from the sky vault.

Diffuse-Porous Wood. Certain hardwoods in which the pores tend to be uniform in size and distribution throughout each annual ring or to decrease in size slightly and gradually toward the outer border of the ring.

Diffuser. 1. A device for distributing air in a forced air heating/ cooling system; often flush-mounted on a ceiling, it has slats to direct the conditioned air evenly into the room or space; components of the ventilation system that distribute and diffuse air to promote air circulation in the occupied space. 2. A lens in a light fixture to diffuse light.

Diffuser, Suction. See Suction Diffuser.

Diffuse Transmission. Transmitted light that is scattered evenly in all directions as it passes through a material.

Diffusion. The natural mingling of two or more substances to form a mixture or solution.

Digital. A digital instrument shows a measurement directly as a number, such as the LED display on a digital watch; compare with Analog.

Dike. A long wall or embankment built to prevent flooding.

Dilapidation. A building in a state of neglect, disrepair, decay, ruin, or deterioration through neglect or misuse.

Diluent. In lacquer, the volatile portion of vehicle not capable by itself of dissolving nitrocellulose.

Diluents. See Thinners.

Dimension. 1. A measurable extent of any kind, such as width, length, height, or volume. 2. Dimension Lumber, See Dimension Stock.

Dimensional Stability. 1. The ability of a fabric to retain its dimensions in service and wet cleaning. 2. The ability of any material to maintain its essential dimensions in the conditions in which it is used.

Dimensional Stabilization. Measures taken to prevent or reduce warping, swelling. or shrinking.

Dimension Lumber. Lumber with a nominal thickness of from 2 inches up to but not including 5 inches and a nominal width of 2 inches or more; includes joists, planks, rafters, studs, and small timbers; see also Dimension Stock.

Dimension Ratio. The diameter of a pipe divided by the wall thickness; each pipe can have two dimension ratios depending on whether the outside or inside diameter is used; in practice, the outside diameter is used if the standards requirement and manufacturing control are based on this diameter; the inside diameter is used when this measurement is the controlling one.

Dimensions of Concrete Masonry Units. In practice, the first dimension of a CMU represents the thickness, the second, height; and the third, length.

Dimension Stock. Hardwood stock processed to a point where the maximum waste remains at the dimension mill and the maximum utility is delivered to the user; stock of specified thickness, width, and length; according to specifications, it may be solid or glued up, rough or surfaced, semifabricated or completely fabricated; commonly known as Hardwood Dimension Lumber.

Dimer. Formed by chemical combination of two similar molecules.

Diminish. To make or appear less or smaller.

Diminishing Returns. An increase in expenditure, investment, or taxation, beyond a certain point, ceases to produce a proportionate yield.

Diminution in Value. The amount by which the market value of property is reduced by any cause.

Dimmer Switch. A switch which controls the brightness of a light.

Dimple. The depression in the surface of gypsum board caused by a hammer as the nail head is set slightly below the surface or plane of the gypsum board to permit concealment with joint compound.

Diode. A semiconductor allowing the flow of current in one direction only and having two terminals.

Diorite. A granular, crystalline igneous rock commonly of acid plagioclase and hornblende, pyroxene, or biotite.

Dip Coating. Process of finishing an article by immersing it in finishing material.

Dipentene. Solvent made by destructive distillation of pine stumps; is stronger than turpentine.

Dip of a Trap. The lowest portion of the inside top surface of the channel through a plumbing trap; compare with Weir of a Trap.

Direct Burial Cable. Electrical cable that is suitable for burial directly in the earth without being in a conduit.

Direct Burial Conduit. Electric conduit suitable for burial in exterior applications with an outer surface that resists moisture, fungus, and corrosion.

Direct Costs. Costs which may readily be attached or directly related to the unit of production.

Direct Current (DC). An electric current, of constant voltage, flowing in one direction only.

Direct Drive Fan. A fan that is positively connected in line to a motor to operate at the same speed.

Direct Examination. Questions directed to a friendly witness.

Direct Expansion (DX). Interior air cooled by directly passing over an evaporator in which refrigerant is expanding from a fixed reference point.

Direct Expansion Evaporator. Evaporator using either an automatic expansion valve (AEV) or a thermostatic expansion valve (TEV) refrigerant control.

Directional Arrow. A symbol used to denote direction.

Directional Sign. A publicly displayed notice which indicates by use of words or symbols a recommended direction or route of travel.

Direction of Irrigation. Direction of flow of irrigation water; usually at right angles to the supply ditch or pipe.

Direct Job Overhead. A firm's overhead expense that is directly attributable to completing a work in process and that would not be incurred if no projects were under way.

Direct Labor. The actual cost of labor payroll for all jobs worked on during the period covered by the income statement.

Direct-Labor Burden. Includes all payroll taxes, insurance, and employees' benefits which are associated with the labor payroll.

Direct Methods. Techniques of solar heating in which sunlight enters a house through the windows and is absorbed inside.

Direct Nailing. See Face-Nailing.

Directory Board. A panel that is used to display information.

Direct Process. See American Process Zinc Oxide.

Direct Radiation. Solar radiation that comes straight from the sun, casting shadows on a clear day.

Direct Transmission. Light passing through a clear, transparent material.

Disability. 1. A physical or mental impairment that substantially limits one or more of the major life activities of an individual. 2. A record of such an impairment. 3. Being regarded as having such an impairment.

Disappearing Stair. A folding, hinged stairway that is pulled down from the ceiling and commonly hidden in a trap door.

Discharge Valve. A special purpose valve.

Discharging Arch. An arch built over a lintel or header to help relieve the pressure from the wall above the arch; also known as a Relieving Arch or Safety Arch.

Discoloration. Change of color from that which is normal or desired.

Disconnect. An electrical switch.

Disconnecting Means. A device, or group of devices, or other means whereby the conductors of a circuit can be disconnected from their source of supply.

Discount. 1. A deduction in price or the amount of a bill. 2. To sell a note for less than its principal sum; this has the effect of increasing the yield to the buyer.

Discovery. A process in which parties to a lawsuit are required to divulge information.

Dishwasher. An electric appliance for washing dishes.

Disinfect Pipe. The act or process of transmitting an antimicrobial agent through a piping system.

Disinfectant. Chemical (also called bacteriacide) used to destroy germs and bacteria.

Disintegration. Decomposition into constituent elements or fragments.

Disk Sander. An electric sanding tool where the sand paper abrasive is mounted on a rotating disk.

Dismissal. The act of dropping a lawsuit.

Dispensary. A building or place where medicines and medical or dental aid are dispensed.

Dispenser, Tissue. See Tissue Dispenser.

Dispersed. Finely divided or colloidal in nature.

Dispersing Agent. An admixture capable of increasing the fluidity of pastes, mortars, or concretes by reduction of interparticle attraction; material used to aid in holding finely divided matter in dispersed state

Dispersion. 1.Suspension of one substance in another. 2. A suspension of very fine particles in a liquid medium, as in paints.

Displacement. Movement away from a fixed reference point.

Displacement, Piston. See Piston Displacement.

Disposal. 1. The process of disposing of something. 2. Garbage Disposal Unit.

Disposal Field. See Leaching Field.

Disposal Suit. A complete body suit worn by persons entering a hazardous materials area or handling hazardous materials.

Distemper. 1. A water paint in which the pigment is mixed with white or yolk of egg, casein or size and which is used for mural decoration. 2. Broadly any of numerous water-based paints for general, especially for household use.

Distensibility. Ability to be stretched.

Distillate. A condensed product produced by cooling vapors of a material heated sufficiently to drive off part of the material in the form of vapor.

Distillation. The process of vaporizing a liquid with heat and condensing the vapor to a liquid for the purpose of purification, fractionation, and forming new substances.

Distillery. A building or place where alcoholic liquor is distilled.

Distilling Apparatus. Fluid-reclaiming device used to reclaim used refrigerants; reclaiming is usually done by vaporizing and then recondensing the refrigerant.

Distortion. 1. The quality or state of being twisted out of a natural, normal, or original shape or condition. 2. A lack of proportion in an optical image resulting from defects in the optical system. 3. Faulty reproduction of radio sound or a television picture, caused by change in the wave form of the original signal. 4. Pavement distortion is any change of the pavement surface from its original shape.

Distributed Load. A load on a structural member that is evenly applied along its entire length.

Distributing Pipes. The piping in a building that distribute the water received from the Service Pipe.

Distribution. 1. The movement of fresh concrete to its placement point. 2. The circulation of collected heat to living areas from collectors or storage. 3. The movement of any fluid or current to its end point.

Distribution Box. The main feed line terminus of an electrical service to which branch circuits are connected.

Distribution Panel. The main electrical control center, which contains switches or circuit breakers, is connected to the service wires, and delivers current to the various branch circuits.

Distribution Rib. A transverse thickening of a one-way concrete joist structure used to allow the joists to share concentrated loads.

Distyle. A building or structure having two columns in front.

Ditch. A long narrow excavation dug in the earth for foundation walls, sewer systems, irrigation systems, electrical cable lines, and utilities; a trench.

Divider, Hospital. See Hospital Divider.

Dividers. A drafting tool, like a compass, but with two adjustable sharp points; used for marking out and scribing edges.

Divider, Terrazzo. See Terrazzo Divider.

Diving Board. A narrow platform, often equipped with a spring and raised off the pool deck by a ladder, attached to the edge of a pool and extending over the water, used to give divers altitude and lift.

Diving Platform. Rigid stand for diving.

DIY. Do it yourself.

DLH Series. A standard Steel Joist Institute designation for longspan steel joists.

D Line Cracks. See D Cracking.

D Load. A constant load that in structures is due to the mass of the members, the supported structure, and permanent attachments or accessories.

Dock Board. Heavy timber used in the construction of the raised platform used for the loading and unloading of trucks.

Dock Bumper. Thick rubber units placed under loading dock openings to absorb the shock and prevent damage when trucks back in for loading or unloading.

Dock Leveler. A mechanical device at a loading dock that adjusts to accommodate the different heights of truck beds.

Dog. Fits on a steel bar to form a tool for cramping or clamping.

Dog Leg. Any device that is crooked or bent like a dog's hind leg.

Dog's Tooth. Brick so laid that their corners project from the face of the wall.

Dogging Device. A mechanism used on exit devices that fastens the cross bar in the fully depressed position, and also retains the latch bolt or bolts in a retracted position, thus permitting free operation of the door from either side.

Dolerite. Any of various coarse basalts or diabase.

Dolly Varden Siding. Beveled wood siding that is rabbeted on the bottom edge.

Dolomite. A mineral of the double carbonate of lime and magnesia having the general formula $CaCO_3MgCA_3$.

Dolomitic. A type of lime or limestone containing calcium carbonate in combination with up to 50% magnesium carbonate.

Dome. An arch rotated about a vertical axis passing through its crown, or highest point.

Dome-Hat. Sealed metal container for the motor compressor of a refrigerating unit.

Dome Light. An overhead light.

Domestic Marble. Marble which comes from the country or area where the structure in which it used is built

Domestic Well. A water well dug for household use.

Domicile. 1. A place of residence; a home. 2. A person's fixed, permanent, and principal home for legal purposes.

Dominant Color. Color that predominates or is outstanding.

Doodlebug. A device, as a divining rod, used in attempting to locate underground piping and conduits.

Doohickey. A small mechanical object whose name is not known or has been forgotten.

Door. A hinged, sliding, or revolving barrier for closing and opening an entrance to a building, room, or cupboard.

Door, Access. See Access Door.

Door, Aluminum. A glazed door constructed with aluminum stiles and rails.

Door Buck. A door frame.

Door Bumper. Rubber tip device mounted on walls or baseboards that prevent door knobs from marring walls.

Door, Chain Hoist. A door that is operated by a chain placed in a pulley or sheave.

Door Closer. A mechanical device attached to a door to make it close automatically.

Door, Commercial. See Commercial Door.

Door Coordinator. A device used on a pair of doors to ensure that the inactive leaf is permitted to close before the active leaf; necessary when an overlapping astragal is present and exit devices and automatic or self-latching bolts are used with closers on both door leaves.

Door, Flush. See Flush Door.

Door, Folding. See Folding Door.

Door Frame Grout. Grout to fill in the voids where a masonry structure meets a metal door frame.

Door Frame. The members that completely surround a door, made of wood or metal, to which the hinges are attached.

Door Framing. The rough construction of the wall immediately surrounding the area of a door using wood or metal materials.

Door Guard. Guard fabricated of steel components to provide protection over interior face of door or prevent damage to glass and to prevent intrusion.

Door Head. The assembly of parts at the top of a door frame including the frame, stop, casing, shims, and flashing.

Door Header. Horizontal beam placed on vertical jack studs which form the uppermost portion of the framing of a door opening.

Door Holder. A device used to hold a door in an open position.

Door, Hollow Core. See Hollow Core Door.

Door Jamb. The vertical members of a door frame.

Door Knob. A projecting knob for operating a lock.

Door Louver. Slanted fins, either fixed or movable, provided on a door for ventilation.

Door Motor. A mechanical device operated by electric current which opens or closes a door.

Door Opener. An electronic device which when activated opens a door by mechanically unlatching the throw bolt of a doorknob or which opens a door on a pivot point on the hinge side of the door.

Door, Overhead. See Overhead Door.

Door Paint. A specific paint, usually a latex or alkyd enamel, used for its ability to be washed, to withstand cracking and scratch resistance.

Door Pull. A handle by which a door is pulled.

Door Rail Hanger. Structural steel shape used to support large rolling door.

Door Removal. The act or process of removing a door from its hinges.

Door, Revolving. See Revolving Door.

Door, Roll-Up. See Roll-Up Door.

Door Seal. Rubberized material attached to door head, jamb, and bottom of a door to aid in preventing air drafting.

Door, Shower. See Shower Door

Door, Sliding Glass. See Sliding Glass Door.

Door, Solid Core. See Solid Core Door.

Door, Special. See Special Door.

Door Stop. A molding nailed to the faces of the door frame jambs to prevent the door from swinging through.

Door Switch. An electric switch installed in a door jamb which turns on a light in a closet when the door is opened and turns off a light when the door is closed.

Door Threshold. See Threshold, Door.

Door Trim. 1. Wood or metal finishing work, often ornamental, used for covering the joints between door jambs and plaster walls. 2. The locks, knobs and hinges on a door.

Door, Vault. See Vault Door.

Door, Vertical Lift. See Vertical Lift Door.

Dope. A term used by plasterers for mortar additives of any type, such as those used to retard or accelerate set.

Dope Coat. In tile setting, neat cement applied to the setting bed.

Dormer. A projecting roofed housing in a sloping roof.

Dormer Window. A window in a dormer.

Dormitory. A room occupied by more than two guests.

Dormitory Wardrobe. A closet in student housing where clothes are kept; commonly a free-standing cabinet made to match the cabinets of the dormitory room.

Dot. A small lump of plaster placed on a surface between grounds to assist the plasterer in obtaining the proper plaster thickness and aid in aligning the surface; see Screed, 1.

Dote. Wood rot or decay.

Double Back. A webbed backing cemented to the backing of tufted, knitted and some woven carpets as additional reinforcement, to provide greater dimensional stability; also called Scrim Back.

Double-Back. See Double-Up.

Double-Bevel Groove Weld. See Groove Weld.

Double Bullnose. A type of trim with the same convex radius on two opposite sides.

Doubled-Up. See Double-Up

Double Duty Case. Commercial refrigerator in which a part of space is for refrigerated storage and part is equipped with glass windows for display purposes.

Double Glass. Double Glazing.

Double Glazing. Two parallel sheets of glass with an airspace between;also called Dual Glazing or Double Glass.

Double Headed Nail. A nail with two closely-spaced heads to permit easy removal; widely used in concrete formwork as a temporary fastener; also called a Duplex Nail.

Double Hung Window. The most common style of operable window, it has two sashes that slide vertically in parallel tracks; a window with two overlapping sashes that slide vertically in tracks.

Double-J Groove Weld. See Groove Weld.

Double Layer. Two layers of gypsum board; various thicknesses may be combined to improve the fire, sound, or structural characteristics.

Double Nailing. A method of applying gypsum board by using two nails spaced approximately 2 inches apart every 12 inches in the field along the framing member to insure firm contact with the framing.

Double Offset. In piping, two offsets installed in succession in the same line.

Double Oven. An oven with two baking compartments.

Double Plate. The two wood plates on the top of a wood framed wall.

Double Regulation. Regulation of both pot and gun air pressure.

Double Reinforcement. A concrete beam with steel on both sides of the neutral axis to resist tension and compression.

Double Reinforcing. To double-up on certain materials to increase strength and stability.

Double Spread. See Spread.

Double Stud. Two adjoining studs in a wood framed wall.

Double Tee. A precast concrete slab element that resembles the letters TT in cross section.

Double Thickness Flare. Copper, aluminum or steel tubing end which has been formed into two-wall thickness, 37 to 45 degrees bell mouth or flare.

Double-U Groove Weld. See Groove Weld.

Double-Up. A method of plaster placement characterized by application in successive operations with no setting or drying time allowed between coats; also called Double-Back, Doubled-Up, Laid Off, Laid On, or Two Coat Work.

Double-Vee Groove Weld. See Groove Weld.

Double Walled Tank. A tank constructed with two walls for leak protection and to increase strength and stability.

Double Wye. A fitting or branch pipe in plumbing, constructed of either cast or wrought iron, which has two side outlets at any angle except a right angle; usually a 45 degree angle.

Douglas Fir. A tall, evergreen tree that furnishes long straight framing lumber; also known as Red Fir, Douglas Spruce, and Yellow Fir.

Douglas Spruce. Douglas Fir.

Dovecote. A shelter for nesting doves.

Dovetail Anchor Slot. A matching interlocking strip or slot which is used with a dovetail fastener.

Dovetail Joint. A joint in which one piece has dovetail shaped pins or tenons which fit into corresponding openings on the other piece.

Dovetail Saw. Saw used for cutting very accurate joints; similar to a tenon saw but with a thinner blade and finer teeth.

Dowel. 1. A steel bar, which extends into two adjoining portions of a concrete or masonry construction, as at a joint in a pavement slab, so as to connect the portions and transfer shear loads. 2. As used in the construction of column and wall sections, a deformed steel reinforcing bar placed so as to transmit tension or compression as well as shear loads. 3. A steel reinforcing bar that projects from a foundation to tie it to a column, wall, slab, porch, or steps. 4. A short cylindrical headless rod of wood or steel inserted into adjoining members to keep them in alignment and hold them together. 5. A long cylindrical strip of metal or wood to be cut into dowels.

Dowel Pin. Dowel.

Dowel Sleeve. Cap of light metal or cardboard on one end of a steel dowel bar to allow free movement of an expansion joint.

Down Payment. The money paid by the buyer, being the difference between the full price and the borrowings.

Downspout Boot. Device at bottom of downspout to receive water from downspout and provide transition to drainage piping.

Downspout Bracket. A metal bracket for securing a downspout to a wall.

Downspout. Vertical pipe usually made from sheet metal or plastic which carries water from the roof gutters to the ground or a storm drain; also called a Leader.

Down Time. The time during which equipment is out of service for essential maintenance or emergency repairs.

Dozer. A term used in the trade for a bulldozer.

Dozer Excavation. The digging out of large areas of earth with a bulldozer.

Drafter. Draftsman.

Draft Gauge. Instrument used to measure air movement by measuring air pressure differences.

Draft Indicator. Instrument used to indicate or measure chimney draft or combustion gas movement; draft is measured in units of .10 in. of water column.

Drafting Board. A flat surface used by drafters on which to produce drawings.

Drafting Machine. A mechanical device with adjustable straight edges that may be clamped onto a drawing board for use by a drafter.

Draft Regulator. Device which maintains a desired draft in a combustion-heated appliance by automatically controlling the chimney draft to the desired value.

Draftsman. 1. One who prepares architectural, construction, or engineering drawings; draftsman. 2. A person who drafts documents.

Drag. In welding, the distance between the point of exit of the cutting oxygen stream and the projection on the exit surface of the point of entrance.

Dragging. See Floating.

Dragline. A bucket attachment for a crane commonly used in a marsh or marine area, that digs soft materials that must be excavated at some distance from the crane, and draws the bucket towards itself using a cable.

Dragon's Blood. A red gum exuded from the fruit of a species of palm; used for coloring varnish; the color is not permanent.

Drag Strut. A structural member used to transfer lateral forces from one vertical element to another.

Drain. 1. A trench, ditch, or pipe designed to carry away waste water. 2. A device that allows for the flow of water from a roof area.

Drainage. 1. The process or means of draining a building or land area of atmospheric moisture or sewage. 2. A system of drains, artificial or natural.

Drainage Head. The farthest or highest spots in a drainage area.

Drainage Pipe. Any pipe in a plumbing system which removes waste water, rainwater, or sewage.

Drainage System. All drain and waste pipes that carry water and waste away from the fixtures.

Drainage, Underslab. The process of continuous interception and removing of ground water from under a concrete slab with the installation of crushed stone and perforated pipe.

Drain Board. A countertop adjoining a sink.

Drain Field. A system of trenches filled with sand, gravel or crushed stone, and a series of pipes to distribute septic tank effluent into the surrounding soil.

Drain, Intercepting (Curtain Drain). A drain that intercepts and diverts ground water before it reaches the area to be protected.

Drain, Intercepting. See Intercepting Drain.

Drain Pipe. Any pipe that carries waste or water-borne wastes in a building drainage system.

Drain Tile. Short-length sections of burned clay, plastic, or concrete pipe, laid with open joints and surrounded with aggregate and covered with asphaltic paper or straw to drain the water from an area.

Drain, Waste and Vent (DWV) System. The assemblage of pipes which facilitates the removal of liquid and solid wastes as well as the dissipation of sewer gases.

Drape. A window covering; curtain.

Drapery. Fabric hangings for use as a curtain.

Drapery Rod. A wall-mounted wood or metal shaft for supporting drapery.

Drapery Track. A U-shaped device mounted to a door or window header, ceiling or floor, used as a guide for draperies.

Draughtsman. Draftsman; drafter.

Draw. A small valley or a gully.

Draw Board. A board fitted in a kitchen cabinet that may be pulled out to create a supplementary work surface, like a breadboard.

Drawboard Joint. A mortise-and-tenon joint with holes so bored that when a pin is driven through, the joint becomes tighter.

Drawbridge. A bridge over water, hinged at one end so that it may be raised or lowered to allow or prevent passage of ships.

Drawer. A sliding boxlike storage compartment, usually part of a cabinet.

Drawer Knob. A cabinet knob.

Drawer Pull. A handle with which to withdraw a drawer.

Drawer Roller. A device used to ease the sliding of a drawer open or shut, usually with a metal or fiber wheel rotating in a metal frame.

Drawer Slides. A mechanism employing guides and rollers that guide and support the drawer, permitting easy operation.

Drawing Room. A formal reception room; a withdrawing room.

Drawings. Graphic representations of buildings and their parts; plans or blueprints.

Drawings, Shop. Drawings of specific items of the project provided by subcontractors or fabricators.

Drawn Glass. Glass sheet pulled directly from a container of molten glass.

Drayage. The charge for transportation or delivery of goods.

Dredge. Deepen a waterway by use of a dredging machine, usually mounted on a barge.

Dressed and Matched (D&M). Tongued and grooved; also called Center Matched; see Tongue and Groove.

Dressed Size. The dimensions of lumber after being surfaced with a planing machine; the dressed size is usually 1/2 to 3/4 inch less than the nominal or rough size; a 2- by 4-inch stud, for example, actually measures about 1-1/2 by 3-1/2 inches.

Dressing Room. 1. In a home, a room for storage of clothes and for dressing. 2. In a theater, a room for storing and changing costumes and for applying and removing make-up.

Drier. 1. Substance or device used to remove moisture from a refrigeration system. 2. A solution added to drying oils in paint to quicken the drying. 3. Composition of certain metals that accelerate drying action of oil when added to paint or varnish; some driers are in dry form, others in paste form; most are solutions of metallic soaps in oils and volatile solvents; they are known as driers, oil driers, Japan driers, liquid driers, and Japans; the metallic soaps most commonly used are those of lead, manganese and cobalt.

Drift. 1. Lateral deflection of a building caused by wind or earthquake loads; the horizontal displacement or movement of structure when subjected to lateral forces. 2. Spray loss, in using a spray gun; also called Overspray.

Drift Pin. A tapered steel shaft used to align bolt holes in steel connections during erection.

Drill. Common name for small boring bit and for the electric drill.

Drilled Well. The act or process of using a rotary drill to dig for water.

Drilling Rock. The act or process of boring holes into rock.

Drip. 1.A painting defect where the paint has started to run. 2. A projecting piece of material so shaped as to throw off water and prevent its running down the face of the wall or other surface of which it is a part.

Drip Cap. A piece of metal or wood trim which is placed on the top of an exterior window header to shed off rain.

Drip Edge. A discontinuity or strip installed at roof eaves or over a window or wall component to force adhering drops of water to fall free of the face of the building rather than run toward the interior; the projection of a window head member or water table to allow the water to drip clear of the side of the building below it.

Drip Loop. A low spot in an aerial wire to allow atmospheric moisture and rainwater to drip off before the wire enters the building.

Drippage. Leakage of roofing bitumen through cracks in the roof deck or over the edges of the roof.

Drip Pan. Pan-shaped panel or trough used to collect condensation from evaporator.

Drive Screw. A screw which is driven in with a hammer but removed by a screwdriver.

Drive-Up Window. An opening in a wall through which transactions can be made with persons in motor vehicles.

Driveway. A passageway for automobiles on private property.

Driveway Apron. A sloping transition from the public street to a private driveway.

Driving Cap. A steel cap placed on the top of a pile for protection from damage during driving.

Dropback. Softening Point Drift.

Drop Black. Bone Black.

Drop Cloth. A canvas, plastic, or paper protective sheet used by painters to cover floors and furniture.

Drop Match. See Set Match.

Drop Panel. A thickening of a two-way concrete floor structure at the head of a column.

Drum Trap. A cylindrical plumbing trap, closed at the bottom, with a cover for cleaning; the pipe from the fixture enters the trap near the bottom; the space between where the pipe enters and the bottom is a settling area; the outlet to the waste pipe is higher up in the trap.

Dry Air. Air that contains no water vapor.

Dry Bulb (DB) Temperature. The temperature of air as registered by an ordinary thermometer; compare with Wet Bulb Temperature.

Dry Bulb Thermometer. An ordinary thermometer for measuring ambient air temperature; compare with Wet Bulb Thermometer.

Dry Capacitor Condenser. Electrical device made of dry metal and dry insulation; used to store electrons.

Dry Cell Battery. Electrical device used to provide DC electricity, having no liquid in the cells.

Dry Cleaner. An apparatus for the washing of fabrics which uses substantially non-aqueous organic solvents.

Dry Edging. Rough edges and corners of glazed ceramic ware due to insufficient glaze coating.

Dryer Receptacle. An electrical outlet wired to match the specific electric current needs and configuration of a clothes dryer.

Dryer, Coin. See Coin Dryer.

Dryer, Darkroom. See Darkroom Dryer.

Dryer. An apparatus that will dry the aggregates and heat them to the specified temperatures.

Dry Film Thickness. Depth of applied coating when dry, expressed in mils.

Dry Ice. Refrigerating substance made of solid carbon dioxide which changes directly from a solid to a gas (sublimates). Its subliming temperature is -109 F. (-78 C.).

Drying. Act of changing from liquid to solid state by evaporation of uncombined water and volatile thinners and by oxidation of oils.

Drying Oil. An oil which, when a thin film of it is exposed to the air, takes on oxygen and becomes hard, tough, and elastic;

drying oils are used in the manufacture of paints and varnishes; linseed oil is a common drying oil.

Drying Shrinkage. A decrease in the volume of concrete upon drying.

Drying Time. Time interval between application and final cure.

Dry Kiln. A chamber having controlled air-flow, temperature, and relative humidity for drying lumber, veneer, and other wood products.

Dry Lap. The absence of bitumen at the overlap of two roofing felts.

Dry Location. A location not normally subjected to dampness or wetness; a location classified as dry may be temporarily subject to dampness or wetness, as in the case of a building under construction.

Dry Mix. A concrete or mortar mix in which there is little or no water.

Dryness. 1. The condition of being free of liquid, especially water. 2. The degree of free moisture contained within a gypsum board product.

Dry Oil. See Dehydrated Oil.

Dry-Out. A condition occasionally occurring in gypsum plaster work which by excessive evaporation or suction has lost some or all of the water necessary for crystallization; appears as a light colored, friable area.

Dry Pack. A mixture of cement and fine aggregate with enough moisture for hydration but dry enough to be rammed into place.

Dry Packing. Placing of zero slump, or near zero slump, concrete, mortar, or grout by ramming into a confined space.

Dry Press Brick. Brick formed in molds under high pressures from relatively dry clay, 5 to 7 percent moisture content.

Dry Rodded Weight. The weight of dry aggregate rodded into a cylindrical container of diameter approximately equal to the height, each of 3 layers rodded 25 times, and the excess aggregate struck off level with the top of the container.

Dry Rot. A term loosely applied to any dry, crumbly rot but especially to that which, when in an advanced stage, permits the wood to be crushed easily to a dry powder; the term is actually a misnomer for any decay, since all fungi require considerable moisture for growth.

Dry Set Mortar. A water-retentive hydraulic cement mortar usable with or without sand; when this mortar is used, neither the tile nor walls have to be soaked during the installation process.

Dry Set Tile. Tile set into an adhesive that seems dry but that adheres on contact.

Dry Sheet. A ply of roofing felt mechanically attached to the deck to prevent asphalt or pitch from leaking into the building below; dry sheets are not part of built-up roofing assemblies.

Dry Spots. Small areas on the face of tile which have been insufficiently glazed.

Dry Spray. Overspray or bounce back; sand finish due to spray particle being partially dried before reaching the surface.

Dry Sprinkler System. A fire extinguishing sprinkler system whose pipes remain empty of water until the system is activated.

Dry Standpipe. A fire fighting pipeline in a building that is dry until filled by external connection to a fire hydrant.

Dry Stone Wall. A wall of stone that has been constructed without the use of mortar or concrete in its joints.

Dry System. Refrigeration system which has the evaporator liquid refrigerant mainly in the atomized or droplet condition.

Dry Tape. The application of tape over gypsum board joints with adhesives other than conventional joint compound.

Dry To Handle. 1. A film of paint is dry to handle when it is hardened sufficiently so that it may be handled without being damaged. 2. Time interval between application and ability to pick up without damage.

Dry To Recoat. In painting, time interval between application and ability to receive next coat satisfactorily.

Dry To Touch. 1. A film of paint is dry to touch when it is hardened sufficiently so that it may be touched lightly without any of it adhering to the fingers. 2. Time interval between application and tack-free time.

Dry Type Extinguisher. A fire extinguishing system that discharges a dry powder by means of compressed gas.

Dry Type Spray Booth. A spray booth where an inhibitor has been added to paint to prevent the surface from drying too rapidly, causing wrinkling or cracking.

Dry Type Transformer. A transformer whose coils and core are not immersed in an oil bath.

Dry Vent. A plumbing vent that does not carry any liquids; it acts as a vent only and carries only air.

Drywall. 1. Wall materials that do not involve the use of plaster, such as gypsum wallboard or wood paneling. 2. A gypsum board product, usually gypsum wallboard.

Drywall Removal. The act or process of demolishing and carrying away old drywall systems in a building or structure.

Drywall Studs. Light gauge metal studs that fit into a top and bottom track to support sheets of gypsum wallboard for constructing interior non-bearing partitions.

Drywall Track. The horizontal light gauge metal top and bottom wall members in an interior gypsum wallboard non-bearing partition.

Dry Well. A deep hole, covered and usually lined or filled with rocks, that holds drainage water until it soaks into the ground.

Drywood Termites. See Termite.

DS. Downspout.

DS Glass. Double strength window glass.

Dual Duct Air Handling System. A system similar to a multi-zone system, except that instead of mixing the hot and cold at the air handling unit, separate hot and cold ducts are run to the space to be conditioned, and a terminal mixing box is provided at the space to be conditioned; a constant supply of air is supplied to various zones.

Dual Glazing. Double Glazing.

Dual Vents. A plumbing vent that connects at the point where two fixture drains come together and acts as a back vent for both fixtures' traps.

Duck Tape. Duct Tape, mistakenly or humorously.

Duct. A pipe, tube, or channel carrying air, gas, liquid, or wires.

Duct Bank. A group, series, or tier of round or rectangular metal pipes used to distribute warm or conditioned air throughout a building or structure.

Duct, Bus. See Bus Duct.

Duct, Feeder. See Feeder Duct.

Duct Heater. A heating element in a duct of an air-handling system.

Ductile. 1. In structures, the ability of a material to deform under tensile load. 2. Of metals, capable of being drawn out into wires, rods, or thin shapes.

Ductile Iron Pipe. Iron pipe that is manufactured so as to render it able to be flexed; ductile iron pipe has the non-corrosive qualities of cast iron but is not brittle and has the handling characteristics of steel.

Ductile Moment Resisting Space Frame. A three dimensional structural system without bearing walls and composed of interconnecting members that resist lateral forces with or without the aid of horizontal diaphragms.

Ductility. 1. The quality of being able to deform considerably under load before fracturing; contrasted to the quality of brittleness. 2. The ability of a substance to be drawn out or stretched thin; while ductility is considered an important characteristic of asphalt cements in many applications, the presence or absence of ductility is usually considered more significant than the actual degree of ductility. 3. The property of a material which allows it to be formed or bent without breaking; copper is more ductile than steel.

Duct Insulation. The material installed on the ductwork in an HVAC system, for the reduction of fire hazard, or as thermal insulation.

Duct Shaft. A lined vertical shaft in a building, usually of a fire-rated construction, through which air ducts and piping are run.

Duct Tape. A tough adhesive tape used to assemble and to repair light gauge air ducts.

Duct, Underfloor. See Underfloor Duct.

Ductwork, Metal. The rigid sheet material out of which the ducts of an HVAC system are manufactured, commonly, galvanized sheet steel, aluminum, or stainless steel.

Due Process. Notice, and an opportunity to be heard.

Dull. Term applied to colors that have a neutral or grayed quality.

Dulling. Loss of gloss or sheen.

Dull Rubbing. Act of rubbing a dried film of finishing material to a dull finish, usually with abrasive paper, pumice stone, steel wool, and oil or water.

Dumbwaiter. A small elevator used for conveying food, dishes, or materials from one floor to another in a building.

Dummy Joint. A joint placed strictly for design, in sidewalks and patios.

Dummy Trim. Hardware trim only, without a lock, usually used on the inactive door in a pair of doors.

Dump Fee. The amount charged for dumping jobsite debris in a landfill or dumpsite.

Dumpster. A large, heavy metal container used to hold and haul rubbish.

Dump Truck. A truck with a tiltable body used for transporting and dumping loose materials.

Dumpy Level. An Engineer's Level.

Dunnage. 1. Loose materials used around a cargo to prevent damage. 2. Scrap gypsum board used for protection of a shipping unit of gypsum wallboard.

Dunting. The cracking that occurs in fired ceramic bodies due to thermally induced stresses.

Duplex. 1. A two-family house. 2. A Duplex Apartment.

Duplex Apartment. An apartment having rooms on two floors.

Duplex Box. An electrical box for wiring switches or duplex receptacles.

Duplex Nail. A Double Headed Nail.

Duplex Paper. Wallpaper which consists of two separate papers pasted together; used to create a highly embossed effect.

Duplex Plate. A protective, finished, plate installed over duplex switches or receptacles.

Duplex Receptacle. A double electric outlet.

Durability. 1. Ability to exist for a long time without significant deterioration. 2. The ability of concrete to resist weathering action, chemical attack, and abrasion. 3. The property of an asphalt paving mixture that describes its ability to resist disintegration by weathering and traffic; under weathering are changes in the characteristics of the asphalt, such as oxidation and volatilization, and changes in the pavement and aggregate due to the action of water, including freezing and thawing.

Duress. Compulsion.

Durham. A type of pipe system composed of iron or steel pipe and all of the joints are threaded screw joints; the fittings used in this system are recessed drainage fittings.

Dust Cap. A protective cap placed over a device to protect it from the intrusion of foreign materials.

Dust Free. A film of paint is dust free when dust no longer adheres to it.

Dusting. 1. The application of dry portland cement to a wet floor or deck mortar surface; a pure coat is thus formed by suction of the dry cement. 2. The development of powdered material on the surface of hardened concrete caused by inadequate curing.

Dust Mask. A fabric mask worn to filter out dust and other foreign matter.

Dustproof. So constructed or protected that dust will not interfere with its successful operation.

Dust-Tight. So constructed that dust will not enter the enclosing case.

Dutch Bond. In masonry, a bond having the courses made up alternately of headers and stretchers; same as English Bond.

Dutch Bond. See English Bond.

Dutch Door Bolt. Device for locking together the upper and lower leaves of a Dutch door.

Dutch Door. A door with two separately hinged leaves, one above the other, enabling one to be open while the other stays shut.

Dutchman. 1. A small wood patch in woodwork. 2. A cut tile used as a filler in the run of a wall or floor area. 3. A narrow strip of carpet side- seamed to standard-width broadloom to compensate for unusual offsets and sloping walls but never used as a substitute for good planning, accurate cutting, and proper stretching techniques.

Dutch Metal. Thin leaves of bright brass which are used for overlaying in the same manner in which gold leaf is applied.

Duty. An obligation that is imposed by law or by contract.

Duty, Continuous. See Continuous Duty.

Dwarf Wall. A short wall or partition that does not reach the ceiling.

Dwelling. A place in which people live.

Dwelling Unit. One or more habitable rooms which are intended or designed to be occupied by one family with facilities for living, sleeping, cooking, and eating.

DWV. Drainage, waste, and vent system.

DX. Direct Expansion.

DYE. A material used for dyeing or staining.

Dynamic. Having to do with bodies in motion; implying motion or change of state; for example, an earthquake force is a dynamic force; the opposite of Static.

Dynamic Force. A moving force; F=ma, or force equals mass times acceleration.

Dynamite. A type of explosive; a slang term used by tilesetters when referring to a mortar accelerator.

Dynamo. An electrical power generator.

Dynamometer. Device for measuring power output or power input of a mechanism.

Dyne. A unit of force that, acting on a mass of one gram, increases its velocity by one centimeter per second along the direction that it acts.